# Key Questions in Zoo and Aquarium Studies:
# A Study and Revision Guide

# Key Questions in Zoo and Aquarium Studies: A Study and Revision Guide

**Paul A. Rees** *BSc (Hons), LLM, PhD, CertEd*
Formerly Senior Lecturer
School of Science, Engineering and Environment, University of Salford,
United Kingdom

CABI is a trading name of CAB International

CABI
Nosworthy Way
Wallingford
Oxfordshire OX10 8DE
UK

Tel: +44 (0)1491 832111
Fax: +44 (0)1491 833508
E-mail: info@cabi.org
Website: www.cabi.org

CABI
WeWork
One Lincoln St
24th Floor
Boston, MA 02111
USA

Tel: +1 (617)682-9015
E-mail: cabi-nao@cabi.org

© Paul A. Rees 2021. All rights reserved. No part of this publication may be reproduced in any form or by any means, electronically, mechanically, by photocopying, recording or otherwise, without the prior permission of the copyright owners.

A catalogue record for this book is available from the British Library, London, UK.

References to Internet websites (URLs) were accurate at the time of writing.

ISBN-13: 9781789249002 (paperback)
        9781789249019 (ePDF)
        9781789249026 (ePub)

DOI: 10.1079/9781789249002.0000

Commissioning Editor: Ward Cooper
Editorial Assistant: Emma McCann
Production Editor: James Bishop

Typeset by SPi, Pondicherry, India
Printed and bound in the UK by CPI Group (UK) Ltd, Croydon, CR0 4YY

*For Harry, George and Elliot, who all love visits to the zoo*

An elephant by Elliot Clark (aged 4⅓ years)

# Contents

# About the Author

Paul Rees was a senior lecturer in the School of Science, Engineering and Environment at the University of Salford, United Kingdom, for 22 years until his retirement in 2020. He holds a BSc in Environmental Biology from the University of Liverpool and a PhD in animal ecology and behaviour from the University of Bradford. Paul previously lectured at three Further Education Colleges and a Higher Education College in the United Kingdom, and trained biology teachers at Sokoto College of Education in Nigeria. He has taught from GCE 'O'/GCSE level to MSc level and has been an external examiner for a range of taught programmes, from Higher National Diploma to MSc level, at six British universities. Paul has published papers on mammal behaviour and ecology, wildlife law, and the role of zoos in conservation, along with eight textbooks concerned with ecology, zoo biology, wildlife law and elephants. He has a longstanding interest in zoos and once worked as an elephant keeper. Paul is the author of three other titles in CABI's *Key Questions* series: *Ecology*, *Applied Ecology and Conservation*, and *Biodiversity*

# Preface

It is not all that long ago that most of the people working in zoos were largely unqualified to work with animals apart from, in some cases, possessing a measure of practical experience. The work attracted former farm workers, animal lovers and even the occasional ex-jockey. Formal qualifications in exotic animal husbandry did not exist and, although it has been true for several centuries that professional zoologists have been involved in establishing and managing large zoos, almost anyone can open a zoo. The professionalisation of the work of keepers is a recent development and in the last quarter of a century there has been a considerable expansion in the number of college and university courses aimed at aspiring zookeepers and zoo scientists. In the United Kingdom it is possible to study animal management and husbandry at BTEC National level (at the age of 16 years) in a college of further education and continue this interest to MSc level and beyond at university.

Alongside this professionalisation of the zoo community there has developed an increased demand for textbooks for both students and their tutors working in this field. The science of zoo biology is still a very young field of scientific endeavour. Most of the available academic books of any relevance are aimed at specialist readers and there are few intended for students new to the field.

This book is intended to introduce students – and anyone interested in zoos and aquariums – to the role of zoos in a modern, environmentally-conscious society. It does this by offering the reader the opportunity to answer 600 multiple-choice questions on a wide range of topics including zoo history, enclosure design, aquarium management, animal behaviour and welfare, zoo research, conservation breeding, zoo visitor behaviour, conservation medicine, zoo legislation and many more.

In recent years many college and university courses have been adopting multiple-choice questions (MCQs) as a method of assessment. In my experience students have a positive attitude to this type of assessment and enjoy practising answering MCQs in class. This trend, and the sudden switch to online teaching and learning at many institutions around the world as a result of the social distancing necessary to slow the spread of the COVID-19 pandemic, have created a unique opportunity for books of MCQs to add value to many science courses.

It is not possible to cover every aspect of the academic study of zoos and aquariums in a book of this size, but I hope it will encourage you to find out more about those topics that interest you and improve your overall understanding of the challenges and rewards of studying, breeding and caring for exotic animals living in captive environments.

# Acknowledgements

I am grateful to Ward Cooper (Commissioning Editor) and his colleagues at CABI for their encouragement and support during the production of this book.

I have published some of the figures in this book elsewhere and I am grateful to my publishers for permission to reuse them: Figs 2.15, 4.7 and 9.3 (Elsevier) and Fig. 9.1 (Wiley-Blackwell). The United States Library of Congress made available the images reproduced here as Figs 1.6 and 2.11. Figure 1.7 is in the public domain and is from an engraving by Antoine Aveline. Figure 6.6 is based on a sign in Edinburgh Zoo.

This book was produced during the COVID-19 pandemic by people in the United Kingdom and India who were working under very challenging conditions and I thank them all for their efforts. I have done my best to keep the text free from errors, as have various people involved in the production process. However, any errors that remain are, of course, mine.

Many of the questions here are based on those I set as part of the assessment for the BSc(Hons) in Wildlife Conservation with Zoo Biology at the University of Salford and I am grateful to those students who have unwittingly tested them for me.

Finally, I should like to thank my grandson Elliot for giving me an excuse to visit zoos with my camera at every opportunity, as if an excuse were really needed.

# How to use this book

The questions in each chapter are divided into three levels: foundation, intermediate, and advanced. These levels are not intended to reflect any particular curriculum but rather general levels of difficulty, and should not be taken too seriously. Knowledge of basic facts are dealt with at the foundation level while the intermediate level and advanced levels contain questions involving more obscure facts and concepts. However, there is some variation between chapters as not all of the areas covered lend themselves to this approach. Students are advised to check the syllabuses they are following in detail before relying too much on this book as a preparation for specific exams.

Students are encouraged to complete a whole chapter – or at least a complete section (foundation, intermediate or advanced) – before looking at the answers. This is because the explanations for some answers may assist in selecting the correct answer to subsequent questions, although I have tried to avoid this where possible. The order in which the chapters are attempted does not really matter because each is about a distinct topic. However, within any chapter you are advised to attempt the foundation questions first, followed by the intermediate questions and finally the advanced questions.

Please note that I have used the term 'zoo' to encompass traditional zoological gardens, safari parks, aquariums and other institutions that keep and exhibit animals for educational, scientific, conservation and commercial purposes. Where possible I have used the names of zoos as they were in the historical context of the questions, for example, London Zoo became ZSL London Zoo.

# 1 History of Zoos and Aquariums

This chapter contains questions on the history of zoos and aquariums along with the famous people and animals associated with them.

## Foundation

**1.1f   The earliest zoo is known from a site in**

a. Rome, Italy

b. Hierakonpolis, Egypt

c. Aleppo, Syria

d. Beijing, China

**1.2f   The term 'menagerie' originated in**

a. Spain

b. Belgium

c. Italy

d. France

**1.3f   The term 'zoo' is an abbreviation of**

a. zoological society

b. zoological collection

c. zoological gardens

d. zoological park

© Paul A. Rees 2021. *Key Questions in Zoo and Aquarium Studies: A Study and Revision Guide* (P.A. Rees)
DOI: 10.1079/9781789249002.0001

**1.4f** **The term 'zoo' was first used in print to describe an animal collection when referring to which of the following?**

a. Clifton Zoo, Bristol

b. Regent's Park Zoo, London

c. Edinburgh Zoo

d. Belle Vue Zoo, Manchester

**1.5f** **Which of the following statements about Carl Hagenbeck is false?**

a. He was an animal trader

b. He opened a zoo at Stellingen in Germany

c. He exhibited humans as well as animals

d. He brought the first elephant to a European zoo

**1.6f** **The oldest zoo still in existence is**

a. Tiergarten Schönbrunn (Schönbrunn Zoo)

b. London Zoo (ZSL London Zoo)

c. Zoo de Vincennes (Paris Zoo)

d. Zoo Basel (Basel Zoo)

**1.7f** **Which zoo is famous for being the first to replace metal bars with open enclosures surrounded by moats and other 'invisible barriers' to contain many species, sometimes giving the (false) impression that several species were present in the same enclosure?**

a. Frankfurt Zoo

b. Berlin Zoo

c. Tierpark Hagenbeck

d. Leipzig Zoo

**1.8f** **Who is considered to be the 'father of zoo biology'?**

a. Carl Hagenbeck

b. Heini Hediger

c. Georges Cuvier

d. Konrad Lorenz

**1.9f** ***Knut* was an orphaned animal who attracted international attention and controversy when he was rejected by his mother and subsequently hand reared in the zoo where he was born in 2006. He was**

a. a polar bear at Berlin Zoo

b. a gorilla at London Zoo

c. a chimpanzee at the Bronx Zoo

d. an elephant at San Diego Zoo

**1.10f** **Which of the following chronological sequences accurately reflects the movements of *Jumbo* the African elephant?**

a. Wild in Sudan → P. T. Barnum (USA) → London Zoo (England) → Jardin des Plantes (Paris)

b. Wild in Sudan → London Zoo (England) → P. T. Barnum (USA) → Jardin des Plantes (Paris)

c. Wild in Sudan → Jardin des Plantes (Paris) → London Zoo (England) → P. T. Barnum (USA)

d. Wild in Sudan → Jardin des Plantes (Paris) → P. T. Barnum (USA) → London Zoo (England)

**1.11f** **Which of the following has a dodo as its emblem (Fig. 1.1)?**

a. The Wildlife Conservation Society

b. The World Association of Zoos and Aquariums

c. The Association of Zoos and Aquariums

d. The Durrell Wildlife Conservation Trust

Fig. 1.1.

**1.12f  Which of the following did not own a private zoo?**

    a.  The Colombian drug lord Pablo Escobar

    b.  The American singer Michael Jackson

    c.  The American businessman and politician William Randolph Hearst

    d.  Each of the above owned a private zoo

**1.13f  Which of the following American zoos specialises in exhibiting species that represent the natural history of the area in which it is located?**

    a.  Woodland Park Zoo

    b.  Arizona-Sonora Desert Museum

    c.  Lincoln Park Zoo

    d.  Alaska Zoo

**1.14f** George Wombwell, Thomas Atkins and Gilbert Pidcock were all

    a. travelling menagerie owners in the 18th and 19th centuries in England

    b. city zoo directors in the United States in the 19th century

    c. exotic animal traders in North America in the early 20th century

    d. circus owners in Victorian England

**1.15f** In June 2010 *Morgan* was rescued from shallow waters in the Waddenzee by a dolphinarium in the Netherlands. This fuelled the debate concerning the keeping of cetaceans in captivity. She was subsequently transported to Loro Parque in the Canary Islands in spite of the efforts of the Free Morgan Foundation to have her released to the wild. *Morgan* was

    a. a minke whale (*Balaenoptera acutorostrata*)

    b. a bottlenose dolphin (*Tursiops truncatus*)

    c. a killer whale (*Orcinus orca*)

    d. a beluga whale (*Delphinapterus leucas*)

**1.16f** Which of the following is the oldest zoo in Australia?

    a. Taronga Zoo

    b. Melbourne Zoo

    c. Perth Zoo

    d. Adelaide Zoo

**1.17f** Which of the following statements about animals in ancient Rome is false?

    a. They were kept in cages under the floor of the Colosseum

    b. Chariots were pulled by horses in races in the Circus Maximus

    c. Most Greek city states possessed menageries

    d. It was rare for animals to be killed for public entertainment

**1.18f** The Mappin Terraces were constructed in London Zoo in 1913–14 and were intended to resemble

    a. mountains

    b. tundra

    c. montane forest

    d. a hot desert

**1.19f** The Chinese Government sent giant pandas (*Ailuropoda melanoleuca*) to the rulers of other countries as gifts up until the mid-1980s. This practice has come to be known as

    a. panda statesmanship

    b. panda diplomacy

    c. panda mediation

    d. panda etiquette

**1.20f** In 1946 the Wildfowl and Wetlands Trust (WWT) opened its first reserve – including a captive collection of wildfowl – at

    a. Slimbridge, Gloucestershire, England

    b. Martin Mere, Lancashire, England

    c. Caerlaverock, Dumfries and Galloway, Scotland

    d. Castle Espie, County Down, Northern Ireland

## Intermediate

**1.1i** In the past many zoos had a Pachyderm House. Which of the following species would have been unlikely to be kept in such a building?

    a. African elephant (*Loxodonta africana*)

    b. Cape buffalo (*Syncerus caffer*)

    c. White rhinoceros (*Ceratotherium simum*)

    d. Hippopotamus (*Hippopotamus amphibius*)

**1.2i** Dudley Zoo, in England, is of historical importance because

    a. it has the oldest chimpanzee exhibit in the United Kingdom

    b. it contains the world's largest collection of buildings designed by the Tecton Group

    c. it has the oldest penguin pool in Europe

    d. it was the first zoo in the United Kingdom to breed Asian elephants

**1.3i** Figure 1.2 is a bust of one of the founders of the World Wildlife Fund (now the World Wide Fund for Nature) who had a life-long interest in the conservation of wildfowl. His name is

    a. Roger Tory Peterson

    b. Jeffery Boswall

c. Sir Peter Scott

d. Sir Edward Grey

Fig. 1.2.

**1.4i   The term 'Garden of Intelligence' refers to a menagerie that existed around 3000 years ago in**

a. Peru

b. England

c. France

d. China

**1.5i   Frédéric Cuvier was at one time responsible for the**

a.   Ménagerie du Jardin des Plantes, Paris

b.   Tiergarten Schönbrunn, Vienna

c.   Jardin Zoologique d'Acclimatation, Paris

d.   Bois de Vincennes Zoo, Paris

**1.6i   The Bärengraben is**

a.   a device for catching bears

b.   a zoo in Austria which only keeps bears

c.   a bear pit in Berne, Switzerland, opened in 1857

d.   a type of food given to bears in German zoos

**1.7i   *Man and Animal in the Zoo: Zoo Biology* was first published in 1970 and written by**

a.   Prof. Heini Hediger

b.   Dr Desmond Morris

c.   Dr Ulysses Seal

d.   Dr Hal Markowitz

**1.8i   Where in the United States was the first SeaWorld opened?**

a.   Orlando, Florida

b.   San Diego, California

c.   Monterey, California

d.   Seattle, Washington State

**1.9i   Between 1773 and 1829 the Exeter Exchange in London housed**

a.   an aquarium

b.   a menagerie

c.   an aviary

d.   a circus

**1.10i   Which of the following was the first true zoo in the United States?**

a.   Bronx Zoo, New York

b.   Lincoln Park Zoo, Chicago, Illinois

    c. Cincinnati Zoo, Ohio

    d. Philadelphia Zoo, Pennsylvania

**1.11i** **Which of the following historical figures was not important in the history of the development of modern zoos?**

    a. Dr Konrad Lorenz

    b. Carl Hagenbeck

    c. Decimus Burton

    d. Sir Stamford Raffles

**1.12i** **Match the names of the gorillas in Table 1.1 with the reason they became well-known.**

**Table 1.1**

| | A | B | C | D |
|---|---|---|---|---|
| A male gorilla at Cincinnati Zoo who was shot dead when a child fell into his enclosure | *Harambe* | *Jambo* | *Harambe* | *Guy* |
| An albino gorilla who lived at Barcelona Zoo | *Jambo* | *Snowflake* | *Snowflake* | *Harambe* |
| An old favourite at London Zoo | *Snowflake* | *Guy* | *Guy* | *Jambo* |
| A male gorilla at Jersey Zoo who protected an unconscious child who had fallen into his enclosure | *Guy* | *Harambe* | *Jambo* | *Snowflake* |

a. A

b. B

c. C

d. D

**1.13i** **The names of the American zoo exhibits listed in Table 1.2 include the names of**

a. zoo directors

b. zoo benefactors

c. zoo scientists

d. zookeepers

Table 1.2

| Exhibit | Zoo |
|---|---|
| Polk Penguin Conservation Center | Detroit Zoo, Michigan |
| Calvin T. Corrieri Meerkat Exhibit | Nashville Zoo, Tennessee |
| Hubbard Orangutan Forest | Henry Doorly Zoo, Omaha |
| Downing Gorilla Forest | Sedgwick County Zoo, Kansas |
| Regenstein Wolf Woods | Brookfield Zoo, Illinois |

**1.14i** **Which King of England established a zoo in the Tower of London?**

a. Henry I

b. George I

c. Charles II

d. Edward V

**1.15i** **The person depicted in Fig. 1.3 is**

a. Sir Stamford Raffles

b. Gerald Durrell

c. William Temple Hornaday

d. John Aspinall

Fig. 1.3.

**1.16i  The entrance to Artis Zoo is shown in Fig. 1.4. In which city is it located?**

   a.  Berlin

   b.  Paris

   c.  Amsterdam

   d.  Copenhagen

Fig. 1.4.

**1.17i   The city states of ancient Greece maintained menageries whose animals were described in his book *History of Animals* by**

    a.  Socrates

    b.  Plato

    c.  Hippocrates

    d.  Aristotle

**1.18i   The first permanent cage in London Zoo was erected in 1828 (Fig. 1.5) and is called the**

    a.  Crow's Cage

    b.  Raven's Cage

    c.  Parrot's Cage

    d.  Eagle's Cage

Fig. 1.5.

**1.19i** Which Aztec emperor kept birds, mammals and reptiles in a large garden said to have been tended by more than 600 staff?

a. Montezuma

b. Itzcoatl

c. Tizoc

d. Ahuitzotl

**1.20i** The Tower of London menagerie was originally located at

a. Richmond Park, London

b. St. James's Park, London

c. Regent's Park, London

d. The Royal Park of Woodstock, Oxfordshire

# Advanced

**1.1a** A walled enclosure or preserve in which animals were kept by Persian kings for hunting, and where a wide variety of trees and other plants were grown, was known as

a. a paradeisos

b. an Elysium

c. a Shangri-la

d. an Arcadia

**1.2a    The person shown in Fig. 1.6 is**

a. William Temple Hornaday

b. Phineas T. Barnum

c. Marlin Perkins

d. Carl Hagenbeck

Fig. 1.6.

**1.3a    Which of the following statements about safari parks is false?**

a. Africa USA was the first drive through safari park in the United States

b. Knowsley Safari Park was the first safari park in the United Kingdom

c. The Highland Wildlife Park specialises in keeping past and present Scottish wildlife

d. The circus owner Jimmy Chipperfield was largely responsible for the development of the safari park concept in the United Kingdom

**1.4a    The Elephant and Rhino Pavilion built at London Zoo between 1962 and 1965 was designed by**

a. Berthold Lubetkin and the Tecton Group

b. Sir Hugh Casson and Associates

c. Jones and Jones

d. Jon Coe Design

**1.5a    Which of the following statements about zoos in the United Kingdom is false?**

a. John Aspinall founded Howletts Wild Animal Park

b. Sir Stamford Raffles founded London Zoo

c. George Mottershead founded Colchester Zoo

d. Molly Badham and Natalie Evans founded Twycross Zoo

**1.6a    Members of the Bartlett Society are primarily interested in**

a. zoo design

b. wildlife conservation

c. captive breeding

d. zoo history

**1.7a    The first insect house was opened in 1881 in**

a. London Zoo

b. Bristol Zoo

c. Bronx Zoo

d. Hamburg Zoo

**1.8a    Ota Benga was exhibited at the Bronx Zoo in New York in the early 20th century. He was**

a. a chimpanzee (*Pan troglodytes*)

b. a gorilla (*Gorilla gorilla*)

    c. an orangutan (*Pongo pygmaeus*)

    d. a human (*Homo sapiens*)

**1.9a   William Temple Hornaday was the first director of**

    a. Lincoln Park Zoo, Chicago

    b. Franklin Park Zoo, Boston

    c. Bronx Zoo, New York

    d. Detroit Zoo, Detroit

**1.10a In which zoo did the last captive quagga (*Equus quagga quagga*) die in 1883?**

    a. London Zoo, England

    b. Artis Magistra Zoo, Amsterdam

    c. Berlin Zoo, Germany

    d. Basel Zoo, Switzerland

**1.11a Which Egyptian ruler established a zoo in 1500 BCE?**

    a. Akhenaten

    b. Ramses II

    c. Queen Hatshepsut

    d. Tutankhamun

**1.12a The Shedd Aquarium was opened in 1930 in**

    a. Washington, D.C.

    b. New York

    c. Chicago

    d. San Diego

**1.13a Which of the following made observations of the animals at London Zoo for his book *The Expression of Emotions in Man and Animals*?**

    a. Desmond Morris

    b. Charles Darwin

    c. Abraham Lee Bartlett

    d. Sir Solly Zuckerman

**1.14a** **J. Villepreux-Power, A. Thynne and J. B. Procter were important figures in the history of zoos and aquariums. They were all**

    a. employees of London Zoo

    b. zoo and aquarium directors

    c. women

    d. nutritionists

**1.15a** **The zoo illustrated in Fig. 1.7 is**

    a. Jardin des Plantes, Paris, France

    b. Regent's Park Zoo, London, England

    c. Tiergarten Schönbrunn, Vienna, Austria

    d. Louis XIV's menagerie at Versailles, near Paris, France

Fig. 1.7.

**1.16a** **The Zoological Society of Victoria was founded in Australia in 1857. In 1861, prior to establishing Melbourne Zoo, it was re-named The Acclimatisation Society of Victoria. Its purpose was to**

    a. transport animals to Australia from around the world to see which species could adapt to the Australian environment and be successfully introduced

    b. educate the public about exotic animals

    c. establish captive breeding programmes for endangered species

    d. establish captive breeding programmes for Australian native species

## 1.17a The Snowdon Aviary (Fig. 1.8) is located in

    a. Bristol Zoo

    b. Edinburgh Zoo

    c. ZSL London Zoo

    d. Dudley Zoo

Fig. 1.8.

**1.18a  Match the authors listed in Table 1.3 with the titles of their books about zoos.**

Table 1.3

| Author | A | B | C | D |
|---|---|---|---|---|
| Colin Tudge | A Different Nature | The Management of Wild Mammals in Captivity | Last Animals at the Zoo | Last Animals at the Zoo |
| Lee Crandall | Zoo 2000 | A Different Nature | The Management of Wild Mammals in Captivity | Zoo 2000 |
| David Hancocks | The Management of Wild Mammals in Captivity | Zoo 2000 | A Different Nature | The Management of Wild Mammals in Captivity |
| Jeremy Cherfas | Last Animals at the Zoo | Last Animals at the Zoo | Zoo 2000 | A Different Nature |

    a.  A

    b.  B

    c.  C

    d.  D

**1.19a  Dr Betsy Dresser is famous for her contributions to developments in**

    a.  environmental enrichment

    b.  assisted reproductive technologies

    c.  exotic animal veterinary medicine

    d.  zoo enclosure design

## 1.20a In which zoo did the last individual of each of the species listed in Table 1.4 die?

Table 1.4

| Extinct species | A | B | C | D |
|---|---|---|---|---|
| Thylacine (*Thylacinus cynocephalus*) | Adelaide Zoo | Barcelona Zoo | Moscow Zoo | Beaumaris Zoo, Hobart |
| Passenger pigeon (*Ectopistes migratorius*) | Bronx Zoo | Dublin Zoo | Berlin Zoo | Cincinnati Zoo |
| Polynesian tree snail (*Partula turgida*) | Singapore Zoo | Toronto Zoo | Edinburgh Zoo | London Zoo |

a. A

b. B

c. C

d. D

# 2 Zoo and Exhibit Design

This chapter contains questions on the methods used to contain animals, zoo design and the design of individual animal exhibits. Aquariums and aquatic exhibits are considered in Chapter 3.

## Foundation

**2.1f** The means by which an animal in a zoo is held safely and separated from staff and visitors by barriers is known as

    a. confinement

    b. containment

    c. internment

    d. impoundment

**2.2f** A reinforced pipe barrier is suitable for an enclosure containing

    a. large felids

    b. apes

    c. ostriches

    d. rhinoceroses

**2.3f** Electrified wires are used in some enclosures to keep the animals away from fences, ornamental vegetation and other easily damaged structures. Such a wire is known as a

    a. safety wire

    b. hot wire

© Paul A. Rees 2021. *Key Questions in Zoo and Aquarium Studies: A Study and Revision Guide* (P.A. Rees)
DOI: 10.1079/9781789249002.0002

c. shock wire

d. base wire

**2.4f    In an enclosure used to house hoofstock, an area of concrete adjoining an animal house where the animals can stand when the other parts of the enclosure are waterlogged is called**

a. hard standing

b. a raft

c. a podium

d. hard core

**2.5f    An electric fence should not normally be used**

a. to protect vegetation

b. to prevent animals from climbing trees

c. as a primary barrier

d. at the top of vertical fence barrier

**2.6f    The purpose of fixing silhouettes of birds of prey on large glass barriers in zoos is to**

a. make the glass visible to visitors

b. add decoration to the glass

c. obscure the visitors to reduce the disturbance to the animals

d. prevent wild birds from colliding with the glass

**2.7f    The barrier shown in Fig. 2.1 is best described as a**

a. ha-ha

b. dry moat

c. psychological barrier

d. depressed wall barrier

Fig. 2.1.

**2.8f** Which of the following taxa are often kept behind laminated glass windows in their indoor accommodation to prevent the transmission of zoonoses?

a. Gazelles

b. Equids

c. Primates

d. Mustelids

**2.9f** Which of the following types of barrier should allow visitors an uninterrupted view of the animals in an enclosure?

i. Ha-ha

ii. Dry moat

iii. Zoomesh

iv. Depressed vertical fence

v. Wet moat

a. i, ii, iv, v

b. i, iv and v

c. ii, iv and v

d. i, ii, iii, iv and v

**2.10f** **A fence bounding a tiger enclosure is shown in Fig. 2.2. What is the name of the part of the fence labelled X?**

a. The return

b. The prop

c. The buttress

d. The stanchion

Fig. 2.2.

**2.11f** **In some walk-through exhibits very low barriers (e.g. wooden logs) are placed along the edges of footpaths that do not physically constrain the animals or the visitors. These are known as**

a. stand-off barriers

b. virtual barriers

c. cognitive barriers

d. psychological barriers

**2.12f** **A simulated subterranean exhibit is most suitable for**

a. Patagonian maras (*Dolichotis patagonum*)

b. pygmy tree shrews (*Tupaia minor*)

c. indris (*Indri indri*)

d. naked mole rats (*Heterocephalus glaber*)

**2.13f** **Pinioning is a technique used to contain some**

a. reptiles

b. birds

c. mammals

d. amphibians

**2.14f** **Which of the following species would not be appropriate for a multi-species African savannah exhibit?**

a. Giraffe (*Giraffa camelopardalis*)

b. Plains zebra (*Equus quagga*)

c. Ostrich (*Struthio camelus*)

d. Okapi (*Okapia johnstoni*)

**2.15f** **Some indoor animal accommodation has a floor consisting of wood chips and peat sitting on a filter pad on a concrete floor with a drain below the pad. This floor functions as a biological system and prevents the build-up of disease causing organisms. It is generally referred to as**

a. an ecofloor

b. a biofloor

c. a filtered floor

d. a sanitary floor

**2.16f** **Which of the following statements about walk-through exhibits containing free-ranging lemurs is false?**

a. Small children in prams or pushchairs should not be allowed to enter the enclosure

b. Visitors should not take food into the enclosure

c. Such exhibits should only contain individuals of a single lemur species as the different species are incompatible

d. Exhibits should have a double door entry system

**2.17f**  In a zoo that experiences a temperate climate, animals referred to as 'herps' are most likely to be housed in

    a.  a terrarium

    b.  an aquarium

    c.  an aviary

    d.  a paddock

**2.18f**  When a guillotine door is being opened it moves

    a.  to the left

    b.  to the right

    c.  up

    d.  down

**2.19f**  A reversed lighting schedule is not appropriate for the

    a.  aye-aye (*Daubentonia madagascariensis*)

    b.  Sunda slow loris (*Nycticebus coucang*)

    c.  Senegal bush baby (*Galago senegalensis*)

    d.  golden lion tamarin (*Leontopithecus rosalia*)

**2.20f**  Visitors to the gorilla exhibit shown in Fig. 2.3 are able to observe the animals at close quarters from behind glass, unobtrusive fences and water barriers. An exhibit such as this, where visitors feel as if they are sharing the landscape with the animals, is known as

    a.  an intrusion exhibit

    b.  an immersion exhibit

    c.  a submersion exhibit

    d.  an engagement exhibit

Fig. 2.3.

# Intermediate

**2.1i** **Which of the following was an important figure in the design of zoo exhibits in the United Kingdom?**

    a.  Dr Desmond Morris

    b.  Dr Ulysses Seal

    c.  Berthold Lubetkin

    d.  Dr Hal Markowitz

**2.2i** **In a zoo enclosure, climbing frames, platforms, perches and similar structures are collectively known as**

    a.  props

    b.  furniture

    c.  fixtures

    d.  fittings

### 2.3i Jon Coe is

a. a landscape designer with expertise in zoo design

b. a zoologist with expertise in environmental enrichment

c. an educationalist with expertise in zoo education

d. a psychologist with expertise in visitor circulation

### 2.4i Zoo exhibits may be divided into three types: first, second and third generation. Match the exhibit type with the characteristics listed in Table 2.1.

**Table 2.1**

| Characteristics | A | B | C | D |
|---|---|---|---|---|
| Open enclosures, 'habitats' and exhibits with a high concrete component; barriers consisting of moats and low rails | 2nd | 3rd | 1st | 2nd |
| A diversity of species kept behind metal bars in cages or in pits with little or no enrichment | 3rd | 2nd | 2nd | 1st |
| Exhibits mimic the natural environment; invisible barriers; multiple viewing areas; visitors and animals share the same landscape | 1st | 1st | 3rd | 3rd |

a. A

b. B

c. C

d. D

**2.5i** The exhibit illustrated in Fig. 2.4 is designed so that two species can use different enclosures (K, L, M or N) each day, but never at the same time. This is an example of

   a. a revolving exhibit

   b. a naturalistic exhibit

   c. a rotational exhibit

   d. an immersive exhibit

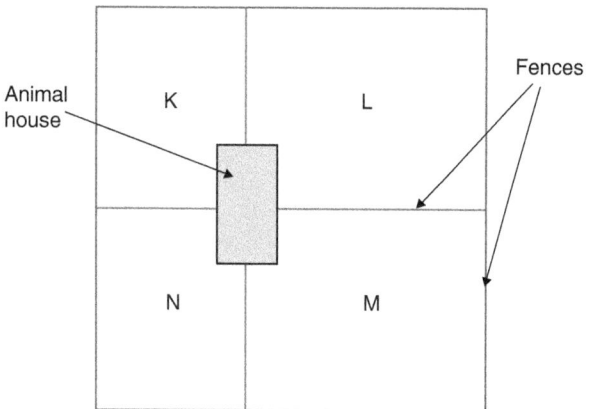

Fig. 2.4.

**2.6i** Double door entry systems are often used in zoos to allow keepers to enter an enclosure without the risk of animal escapes (Fig. 2.5). Which of the following is the correct sequence of events when a keeper enters an enclosure through such a system?

   a. Open outer door; enter space between doors; close outer door; open inner door; enter enclosure; close inner door.

   b. Open outer door; close outer door; enter space between doors; open inner door; enter enclosure; close inner door.

   c. Open outer door; enter space between doors; open inner door; close outer door; enter enclosure; close inner door

   d. Open outer door; enter space between doors; close outer door; open inner door; close inner door; enter enclosure

Fig. 2.5.

**2.7i    The diagram below (Fig. 2.6) shows four possible ways of arranging a stand-off barrier in relation to vegetation and a chain-link fence (A-D). Which of these is most appropriate for the protection of the public?**

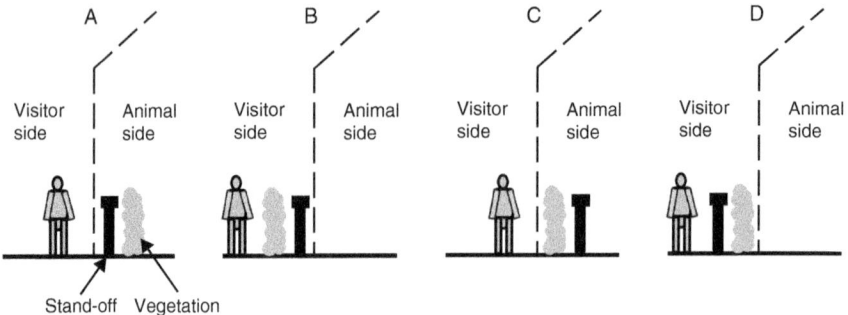

Fig. 2.6.

   a. A

   b. B

   c. C

   d. D

**2.8i** **A transfer chute in the indoor accommodation for lions is used to**

a. transfer food into the cage safely

b. transfer lions safely from one cage to another

c. transfer water to a water trough

d. allow the keepers to transfer from one part of the cage to another safely

**2.9i** **Complete the following sentence using one of the options below: 'In a vivarium containing reptiles ............. should be maintained so that the animals can select their preferred temperature'.**

a. a high temperature

b. an ambient temperature

c. a constant temperature

d. a thermal gradient

**2.10i** **Which of the following organisms are most under-represented in zoos and aquariums compared with the number of species that exist in nature?**

a. Fishes

b. Amphibians

c. Invertebrates

d. Reptiles

**2.11i** **A ha-ha is a containment barrier used in the design of some zoo enclosures. It is based on a device used in landscape design to hide the presence of walls that was used extensively in garden design by**

a. Lancelot 'Capability' Brown

b. Carl Hagenbeck

c. Berthold Lubetkin

d. Sir Stamford Raffles

**2.12i** Woodland Park Zoo in Seattle is widely recognised as having produced the first naturalistic enclosure in the world. The enclosure was designed for

a. tigers

b. gorillas

c. elephants

d. orangutans

**2.13i** The diagram below (Fig. 2.7) shows vertical sections through a chain linked fence indicating the position of the overhang and locations of electric fences (hot wires) at the top and bottom of each fence. Which image (A-D) shows the correct configuration of the fence if climbing animals are to be safely contained?

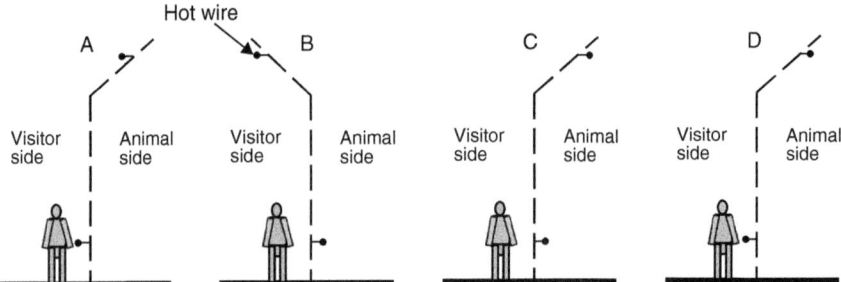

Fig. 2.7.

a. A

b. B

c. C

d. D

**2.14i** Figure 2.8 is a map of a zoo showing the locations of animal houses and paths. Which locations (A-D) are the best and the worst positions for a new exhibit of shy primates that are unlikely to breed successfully if constantly disturbed?

a. Best = A; worst = C

b. Best = B; worst = D

c. Best = C; worst = A

d. Best = C; worst = D

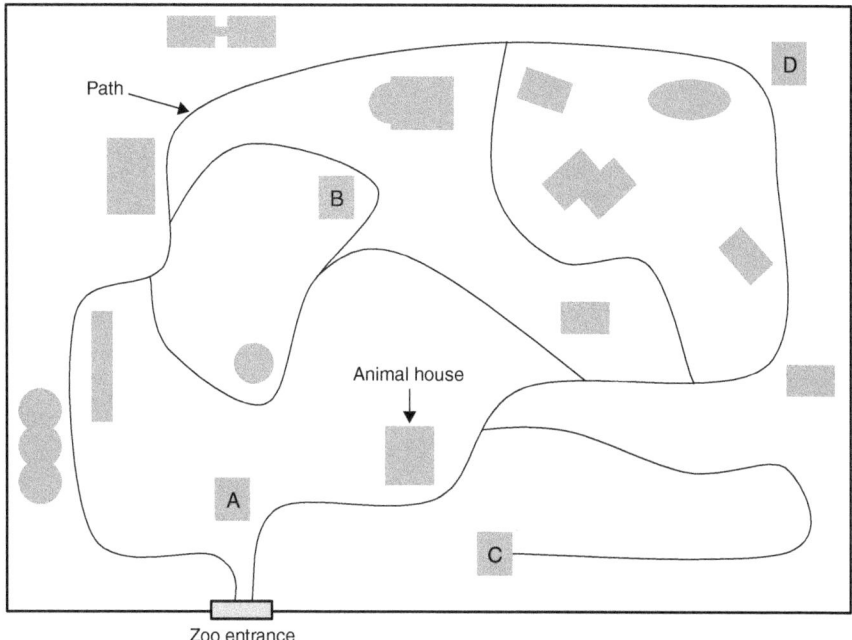

Fig. 2.8.

**2.15i** **A zoo exhibit where visitors and animals are able to share the same setting but not the same space describes the concept of**

    a. landscape immersion

    b. a naturalistic exhibit

    c. an enriched exhibit

    d. landscape realism

**2.16i** **The enclosures designed by Lubetkin were predominantly made of**

    a. natural materials

    b. concrete

    c. brick

    d. wood

**2.17i** **The building shown in Fig. 2.9 is a giraffe house opened in 1836 at**

    a.  Dublin Zoo

    b.  Edinburgh Zoo

    c.  Berlin Zoo

    d.  London Zoo

Fig. 2.9.

**2.18i** **The natural circadian rhythms of animals kept in zoos can be maintained by**

    a.  supplying an appropriate substrate

    b.  providing an appropriate ambient temperature throughout the day

    c.  the careful management of lighting regimes

    d.  suitable adjustments to diet

**2.19i** **A post-occupancy evaluation of a new exhibit may be used to**

    a.  assess the effects of the design on animal welfare

    b.  determine the effectiveness of the interpretation signage in educating zoo visitors

c. inform and improve future designs

d. do all of the above

**2.20i    A puparium in a zoo would contain**

a. butterfly cocoons

b. leaf cutter ants

c. tarantulas

d. scorpions

# Advanced

**2.1a    The Penguin Pool at London Zoo (Fig. 2.10) was completed in 1934 and designed by**

a. Foster + Partners

b. Berthold Lubetkin

c. Jones and Jones

d. Hugh Casson

Fig. 2.10.

2.2a **Zoo360 is an extensive series of elevated interconnected mesh passageways that allow animals such as tigers, apes, monkeys and lemurs to travel around**

a. Taronga Western Plains Zoo

b. ZSL London Zoo

c. Philadelphia Zoo

d. San Diego Zoo

2.3a **The term 'open range zoo' is most closely associated with**

a. American zoos exhibiting rangeland species

b. drive-through British facilities exhibiting African savannah species

c. European zoos containing naturalistic exhibits of temperate grassland species

d. drive-through facilities in Australia

2.4a *Gunite* **is widely used in zoo exhibits to**

a. disinfect indoor accommodation

b. produce simulated rock surfaces

c. replace glass

d. reinforce fences

2.5a **The building in Fig. 2.11 is known as the 'Buffalo House'. It was built in 1891 in the style of a log cabin and is located in the**

a. Bronx Zoo, New York

b. San Francisco Zoo, California

c. Brookfield Zoo, Chicago

d. National Zoological Park, Washington, D.C.

Fig. 2.11.

**2.6a    A paludarium is a type of**

    a.  vivarium

    b.  aquarium

    c.  incubator

    d.  aviary

**2.7a    The 'vertical zoo' is**

    a.  an architectural design concept for a high-rise urban zoo

    b.  a zoo in Singapore

    c.  a combined zoo and aquarium in Saudi Arabia

    d.  a zoo located on a steep hillside in Germany

**2.8a    The period in zoo history when institutions became obsessed with maintaining very high standards of hygiene and sanitation in their cages and enclosures has been referred to as the**

    a.  Sanitation Era

    b.  Antiseptic Era

c. Disinfectant Era

d. Sanitisation Era

**2.9a   In the historical period referred to in Q2.8a, what was the effect of the prevailing attitude towards cleanliness?**

a. Poorer welfare for the animals

b. A poorer experience for visitors

c. Improved welfare for animals

d. Poorer welfare for the animals and a poorer experience for visitors

**2.10a   Animal J has a flight distance of 7 metres and is very sensitive to the presence of visitors. Assuming the animal has access to no other space, and visitors can access the entire perimeter of the enclosure, which of the following enclosure shapes would be unsuitable? (Note Fig. 2.12 shows plan views).**

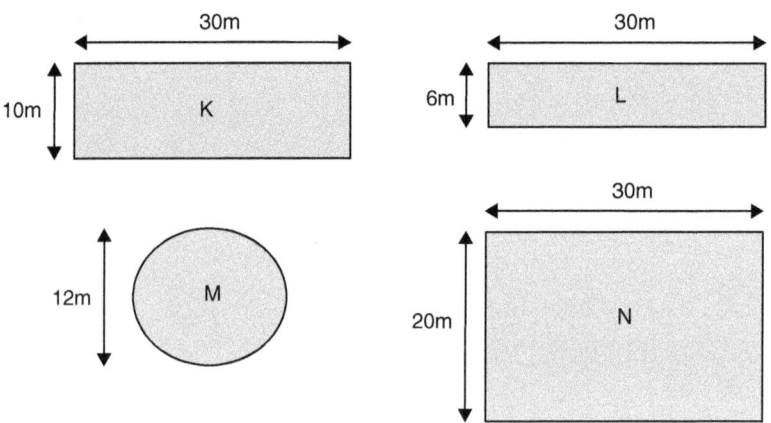

Fig. 2.12.

a. K and L

b. L and M

c. L, M and N

d. K, L and M

2.11a  The space allocations for enclosures for Emperor penguins (*Aptenodytes fosteri*) are shown in Table 2.2 What is the minimum total land area and the minimum total water volume required for an enclosure holding 14 penguins?

Table 2.2

| | Land surface area (m²) | Water surface area (m²) | Pool depth regardless of number (m) |
|---|---|---|---|
| Per bird for the first 6 birds | 1.67 | 1.67 | 1.33 |
| Per bird for each additional bird | 0.84 | 0.84 | |

a.  Land = 16.74m²; water volume = 16.74m³

b.  Land = 16.74m²; water volume = 22.26m³

c.  Land = 13.53m²; water volume = 13.11m³

d.  Land = 19.29m²; water volume = 21.43m³

2.12a  Which of the following were granted a patent in the United States in 1924 for a new type of zoo enclosure containing the following as part of the description of their invention?

*...Another object of the invention is the provision of means for separating and enclosing the animals in such a manner that a panoramic view may be obtained of a part of or the entire exhibition, while the separating and enclosing means remain practically hidden.*

a.  Sir Hugh Casson and Associates

b.  Konrad Lorenz and Niko Tinbergen

c.  Heinrich Hagenbeck and Lorenz Hagenbeck

d.  Berthold Lubetkin and the Tecton Group

2.13a  The concept of an 'inverted zoo' or 'reversed zoo' alters the relationship between the animals and the public by

a.  'caging' the visitors instead of the public

b.  exhibiting all animals in multi-species exhibits

   c. housing predator and prey species in the same enclosure

   d. containing animals behind barriers that are 'invisible' to the public

**2.14a** **The diagrams that make up Fig. 2.13 are plan views of four options for the indoor accommodation for a large group of chimpanzees (*Pan troglodytes*). Which shape is most appropriate if serious aggressive interactions between individuals are to be kept to a minimum?**

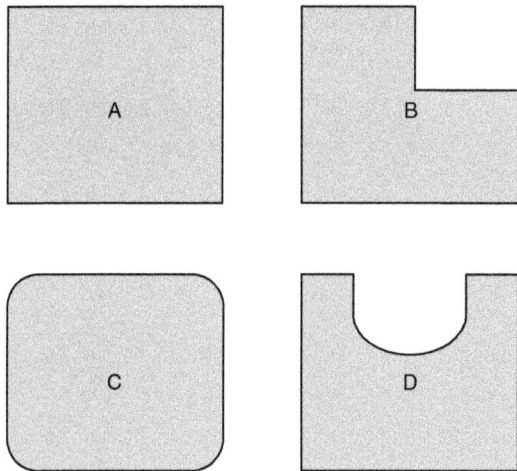

**Fig. 2.13.**

   a. A

   b. B

   c. C

   d. D

**2.15a** **Figure 2.14 shows a vertical section through a fence used to retain a species of large carnivore. The steel mesh extends down the posts and then horizontally away from the posts under the ground on the inside of the enclosure to prevent the animals from digging out at the base. The section of fence labelled 'X' is called the**

   a. base mesh

   b. pedestal

   c. foundation skirt

   d. ground skirt

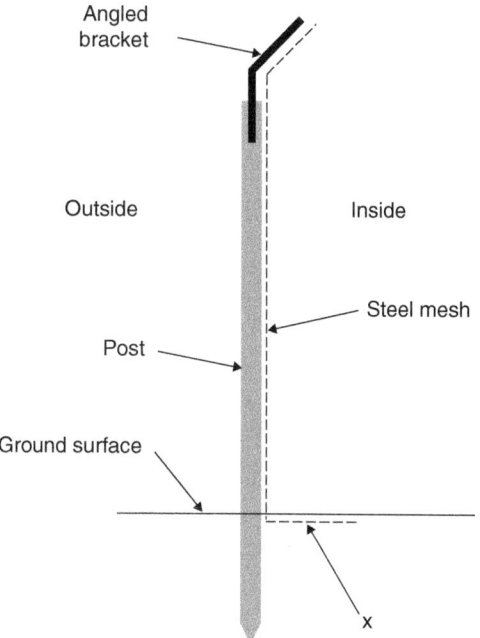

**Fig. 2.14.**

**2.16a** **Which of the following zoo exhibits is most likely to be considered to be encouraging a dominionistic attitude towards animals?**

    a. A flat grass paddock containing giraffes (*Giraffa camelopardalis*) where the visitor viewing is at ground level

    b. An enclosure for Asian elephants (*Elephas maximus*) where they are separated from the public by a ha-ha

    c. A walk-through exhibit containing ring-tailed lemurs (*Lemur catta*)

    d. A bear pit containing brown bears (*Ursus arctos*)

**2.17a** **Match the names of the exhibits in Table 2.3 with the names of the zoos in which they are located.**

**Table 2.3**

| Exhibit | A | B | C | D |
|---|---|---|---|---|
| *Arctic Ring of Life* | Chester Zoo | Detroit Zoo | Detroit Zoo | Hanover Adventure Zoo |
| *Elephant Odyssey* | Hanover Adventure Zoo | San Diego Zoo | Chester Zoo | Detroit Zoo |
| *Spirit of the Jaguar* | Detroit Zoo | Chester Zoo | Hanover Adventure Zoo | San Diego Zoo |
| *Yukon Bay* | San Diego Zoo | Hanover Adventure Zoo | San Diego Zoo | Chester Zoo |

    a. A

    b. B

    c. C

    d. D

**2.18a The system of elephant management that requires keepers to be separated from elephants by a steel fence (Fig. 2.15) is called**

    a. barrier keeping

    b. protected contact

    c. non-contact keeping

    d. barrier contact

Fig. 2.15.

2.19a  **Some exhibits emphasise the interrelationships between wildlife and indigenous peoples (Fig. 2.16) and are examples of**

a.  cultural resonance

b.  cultural dissonance

c.  naturalistic appreciation

d.  indigenous empathy

Fig. 2.16.

2.20a  **When designing a terrarium for a lizard species the required ambient temperature should be determined from knowledge of**

a.  its latitudinal range only

b.  its elevational range only

c.  its latitudinal range and its elevational range

d.  its historical range and its latitudinal range

# 3  Aquariums and Aquatic Exhibits

This chapter contains questions on the design, construction and operation of aquariums and aquatic exhibits and the management of water quality.

## Foundation

**3.1f**  **Which of the following peoples were the earliest aquarists?**

a.  Romans

b.  Ancient Egyptians

c.  Sumerians

d.  Chinese

**3.2f**  **Which zoo opened the first public aquarium (The Fish House)?**

a.  London Zoo, England

b.  Berlin Zoo, Germany

c.  Ménagerie du Jardin des Plantes, Paris

d.  Tiergarten Schönbrunn, Vienna

**3.3f**  **Which of the following is not an aquarium in the United Kingdom?**

a.  Blue Planet

b.  The Deep

© Paul A. Rees 2021. *Key Questions in Zoo and Aquarium Studies: A Study and Revision Guide* (P.A. Rees)
DOI: 10.1079/9781789249002.0003

c. Deep Sea World

d. Two Oceans Aquarium

**3.4f**  **Which of the following is used as a disinfectant in aquariums?**

a. $H_2S$

b. $O_3$

c. $SO_2$

d. $CH_4$

**3.5f**  **If animal faeces accumulate in stagnant water they are likely to cause**

a. eutrophication

b. denitrification

c. condensation

d. oxidisation

**3.6f**  **A generic schematic diagram of a basic aquatic life support system is illustrated in Fig. 3.1. Match the types of filtration listed in the columns in Table 3.1 with the elements labelled K, L and M in the diagram.**

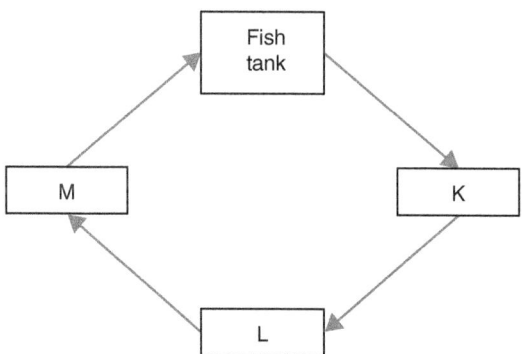

Fig. 3.1.

Table 3.1

|   | A | B | C | D |
|---|---|---|---|---|
| K | Chemical | Mechanical | Mechanical | Biological |
| L | Mechanical | Biological | Chemical | Chemical |
| M | Biological | Chemical | Biological | Mechanical |

a. A

b. B

c. C

d. D

**3.7f** **The generally accepted standard salinity of natural seawater –**
**in parts per thousand (ppt) – is**

a. 15ppt

b. 35ppt

c. 75ppt

d. 110ppt

**3.8f** **An aquarist recorded the pH of the water in five independent**
**aquarium tanks and tabulated the results (Table 3.2). Which**
**of these tanks contained water that is alkaline?**

Table 3.2

| Tank number | 1 | 2 | 3 | 4 | 5 |
|---|---|---|---|---|---|
| pH | 6.8 | 7.0 | 7.9 | 7.1 | 6.7 |

a. 1

b. 1 and 5

c. 3 and 4

d. 1, 2 and 5

**3.9f** **Who opened an aquarium in 1856 in the American Museum**
**in New York City?**

a. William Temple Hornaday

b. Robert Lacy

c. Henry Doorly

d. Phineas T. Barnum

**3.10f** **In 1938 the first oceanarium in the world was opened in**

a. California

b. Florida

    c. New Jersey

    d. Hawaii

**3.11f The specific gravity of aquarium water can be measured using a**

    a. hydrometer

    b. hygrometer

    c. potentiometer

    d. ammeter

**3.12f In a fluidised bed filter the 'bed' consists of**

    a. sand

    b. silica chips

    c. plastic pellets

    d. any of the above

**3.13f With respect to aquariums, cycling time refers to**

    a. the time taken for a single molecule of water to make a complete circuit of the tank

    b. the time taken for a single molecule of water to pass through a biological filter

    c. the time between the first filling of an aquarium with water and when the water should be completely replaced

    d. the time taken for a new aquarium to build up sufficient nitrifying bacteria to convert ammonia and nitrite to nitrate efficiently

**3.14f For most tropical fish species the optimum water temperature range is**

    a. 15–21°C (59–70°F)

    b. 22–28°C (72–82°F)

    c. 29–31°C (84–88°F)

    d. 32–35°C (90–95°F)

**3.15f** Care must be taken to provide adequate structural support for aquarium tanks because water is heavy. One litre of water weighs 1kg. How much would the water weigh that completely filled a tank of dimensions 150cm x 50cm x 80cm?

    a. 510kg

    b. 560kg

    c. 600kg

    d. 650kg

**3.16f** In an aquarium a 'pig' might be used to

    a. clean wide pipes that carry seawater to the aquarium

    b. filter water obtained from the sea

    c. pump water from a sump to the main tank

    d. remove chemical pollutants from water

**3.17f** Which of the following could not be a function of a refugium in an aquarium?

    a. Denitrification

    b. Plankton production

    c. Food animal containment (e.g. copepods and amphipods)

    d. All of the above could be functions of a refugium

**3.18f** In an aquarium a refractometer may be used to measure

    a. evaporative loss

    b. salinity

    c. calcium concentration

    d. water temperature

**3.19f** The process by which the biological filter in an aquarium develops populations of bacteria sufficient to remove all of the ammonia and nitrite produced by the resident fishes and other animals is known as

    a. gestation

    b. mellowing

c.  suppuration

d.  maturation

**3.20f  In an aquarium, polymethyl methacrylate is used**

a.  as a food supplement

b.  to remove chlorine from water

c.  to make windows and tunnels

d.  to treat fish parasites

# Intermediate

**3.1i  Which of the following is most likely to be equipped with a calcium reactor?**

a.  A reef aquarium

b.  A dolphinarium

c.  A vivarium

d.  A terrarium

**3.2i  Which of the following could not be added to a marine pool containing cetaceans to increase the pH?**

a.  Caustic soda

b.  Sodium hypochlorite

c.  Sodium bicarbonate

d.  Carbon dioxide

**3.3i  Some water filters may be cleaned by reversing the flow of water through them so that impurities do not build up on the filter media. This process is called**

a.  through washing

b.  backwashing

c.  flow reversal

d.  reverse cleansing

**3.4i  The drum in a rotary vacuum filter rotates so that**

a.  the water is agitated

b.  the water is aerated

c. the water flow rate through the filter is maintained

d. solids can be pulled out of the liquid being filtered

**3.5i** **A panel used to restrain, regulate or deflect the flow of water in an aquarium system is known as a**

a. baffle

b. bar

c. foil

d. filter

**3.6i** **In an aquarium, what is the purpose of the small plastic balls shown in Fig. 3.2?**

a. They act as enrichment for fish

b. They act as a mechanical filter to remove large particles

c. They provide a large surface area over which a biofilm may form in a biological filter

d. They provide safe places for fish to lay their eggs

Fig. 3.2.

**3.7i** **A foam fractionator in an aquarium is also known as a**

a. calcium reactor

b. protein skimmer

c. froth reactor

d. foam remover

**3.8i** Aquariums may have one of three types of water systems: open, closed or semi-closed. Which of the following statements about aquarium water systems is false?

    a. The pipes in open systems are unlikely to be affected by biofouling

    b. A semi-closed system is an open system that is capable of recirculating water if the water source becomes unreliable

    c. A closed system is most likely to be used in an aquarium located far from the sea

    d. Most closed systems use artificial seawater

**3.9i** The turbidity of water is a measure of its clarity or optical transmission. It depends on the presence of

    a. particles

    b. dissolved oxygen and entrained gases

    c. ozone and other chemicals

    d. all of the above

**3.10i** When analysing marine water, what units are used in the measurement of specific gravity?

    a. $kg/m^3$

    b. ppt

    c. ppm

    d. None

**3.11i** What wavelengths of light are used to kill microorganisms in an aquarium?

    a. Red

    b. Infrared

    c. Ultraviolet

    d. Blue

**3.12i** What type of device is illustrated in Fig. 3.3? The device is shown with the water pump switched off (A) and switched on (B).

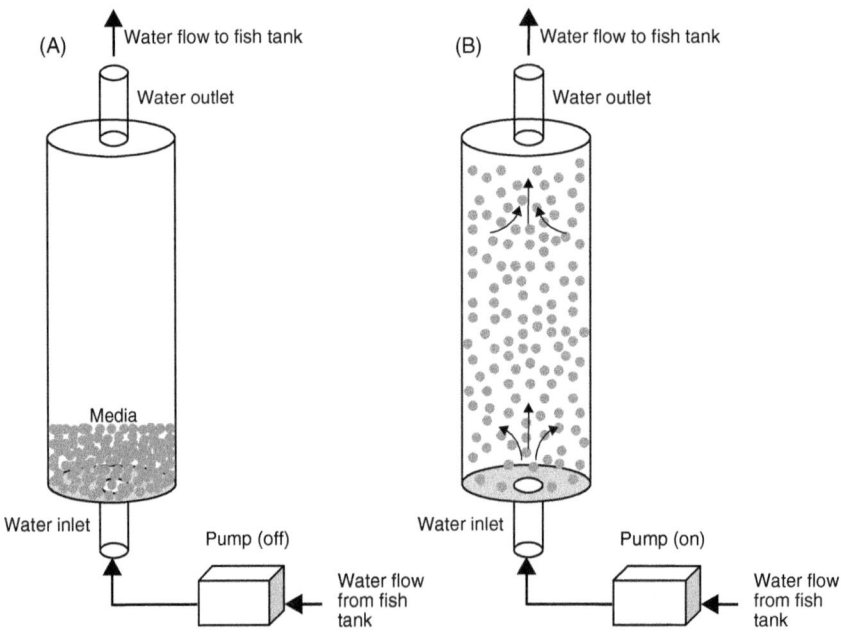

Fig. 3.3.

 a. Trickling filter

 b. Fluidised bed filter

 c. Deaeration tower

 d. Drum filter

**3.13i In an aquarium the redox potential of the water is a measure of**

 a. water quality

 b. water turbidity

 c. water temperature

 d. water pH

**3.14i Which of the following statements about water hardness is false?**

 a. Hardness is made up of temporary and permanent hardness

 b. Temporary hardness is also called carbonate hardness

 c. Permanent hardness is also called non-carbonate hardness

 d. Hardness is entirely the result of the presence of calcium minerals in the water

**3.15i** **If 2 litres of water are lost by evaporation from a small marine aquarium it should be topped up with**

    a.  2 litres of salt water

    b.  2 litres of fresh water

    c.  1 litre of salt water and 1 litre of fresh water

    d.  1 litre of salt water

**3.16i** **The specific gravity of the water in a marine aquarium kept at 24° – 25°C (75 – 77°F) should be in the range**

    a.  1.010–1.013

    b.  1.015–1.019

    c.  1.021–1.024

    d.  1.026–1.030

**3.17i** **Water in a new aquarium tank may be seeded with bacteria from a healthy established tank to avoid**

    a.  eutrophication

    b.  new tank syndrome

    c.  denitrification

    d.  maturation

**3.18i** **In an aquarium the total volume of water in circulation can be increased by**

    a.  adding a larger pump

    b.  increasing the water circulation rate

    c.  adding a sump

    d.  adding a baffle

**3.19i** **Which of the following is unlikely to be found in a sump?**

    a.  A protein skimmer

    b.  A calcium reactor

    c.  A heater

    d.  Fish

3.20i In an aquarium, a shallow pool where visitors are allowed briefly to handle marine organisms such as starfishes under the supervision of a member of staff is known as a

    a. handling pool

    b. touch pool

    c. inspection pool

    d. interpretation pool

# Advanced

3.1a Activated charcoal is used as a chemical filter in some aquariums. It removes chemicals by the process of

    a. adsorption

    b. advection

    c. adhesion

    d. absorption

3.2a Levels of which of the following need to be closely monitored and maintained for corals to thrive in a marine tank?

    a. Calcium and magnesium

    b. Calcium and alkalinity

    c. Magnesium and alkalinity

    d. Calcium, magnesium and alkalinity

3.3a Which of the following sequences occurs in a biological filter?

    a. ammonia → nitrite → nitrate

    b. nitrite → nitrate → ammonia

    c. nitrate → nitrite → ammonia

    d. ammonia → nitrate → nitrite

3.4a A reverse osmosis unit

    a. filters water using centrifugal force

    b. removes contaminants from water using a semi-permeable membrane

c. filters water through a layer of foam sponge using gravity

d. cleans water using a counter-current exchange mechanism

**3.5a** **In a closed aquarium system containing artificial seawater for cetaceans which of the following is not required?**

a. Recirculation through filters

b. Recirculation through fractionators

c. Chemical treatment

d. The addition of salt to offset the effect of evaporation and other losses

**3.6a** **In which of the following locations would you not expect to find a sump?**

a. Above the main tank

b. Alongside the main tank

c. Inside the main tank

d. Underneath the main tank

**3.7a** **The dissolved oxygen concentration in water**

a. increases as temperature increases

b. decreases as temperature increases

c. is unaffected by temperature

d. is positively correlated with temperature

**3.8a** **A spike in the ammonia or nitrite in the water of a mature aquarium tank may be caused by which of the following events?**

i. Filter failure

ii. Excessive cleaning of biological media

iii. Addition of too many fish at the same time

iv. Overfeeding of fish

v. Excessive use of medication

a. i, ii and iii

b. ii, iii and iv

c. i, iii, iv and v

d. i, ii, iii, iv and v

**3.9a   The purpose of a heat exchanger in an aquarium is to**

a. heat cold water

b. cool hot water

c. heat cold water or cool hot water

d. maintain a constant oxygen concentration in the water

**3.10a  Water flow in an aquarium designed for jellyfish is most likely to be**

a. fast and linear

b. slow and linear

c. fast and circular

d. slow and circular

**3.11a  Phosphate and silicate may be removed from freshwater and saltwater aquariums by**

a. an adsorption granulate

b. an absorption granulate

c. an adhesion granulate

d. an advection granulate

**3.12a  In aquarium systems the action of ozone is most accurately described as a**

a. fungicidal

b. biocidal

c. bacteriocidal

d. virucidal

**3.13a  The redox potential of water is measured in**

a. millivolts

b. milliamps

c. millisieverts

d. millimoles

**3.14a  A MultiCyclone is**

    a.  a type of trickling filter

    b.  a centrifugal pre-filtration device

    c.  a fluidised bed reactor

    d.  a type of ion exchange filter

**3.15a  To what do organic compounds bind in a protein skimmer?**

    a.  Charcoal

    b.  A foam pad

    c.  Bioballs

    d.  Small bubbles

**3.16a  The term 'aquarium' was first used in its modern sense by**

    a.  Abraham Dee Bartlett

    b.  Philip Henry Gosse

    c.  Henry Doorly

    d.  Félix Dujardin

**3.17a  The American documentary *Blackfish* (2013) was instrumental in changing the law in relation to the welfare requirements that must be considered by anyone keeping cetaceans in the United States. The main subject of the documentary was *Tilikum*,**

    a.  a bottlenose dolphin (*Tursiops* sp.)

    b.  a killer whale (*Orcinus orca*)

    c.  a beluga whale (*Delphinapterus leucas*)

    d.  a false killer whale (*Pseudorca crassidens*)

**3.18a  The global conservation and sustainability strategy for aquariums published by WAZA in 2009 was called**

    a.  *Ocean Rescue*

    b.  *Ocean Conservation*

    c.  *Turning the Tide*

    d.  *A Future for Fishes*

**3.19a** The changes in the concentration of three nitrogenous compounds with time in the water of a newly-established aquarium containing no fish are illustrated in Fig. 3.4. Match the graph lines to the correct compounds listed in Table 3.3

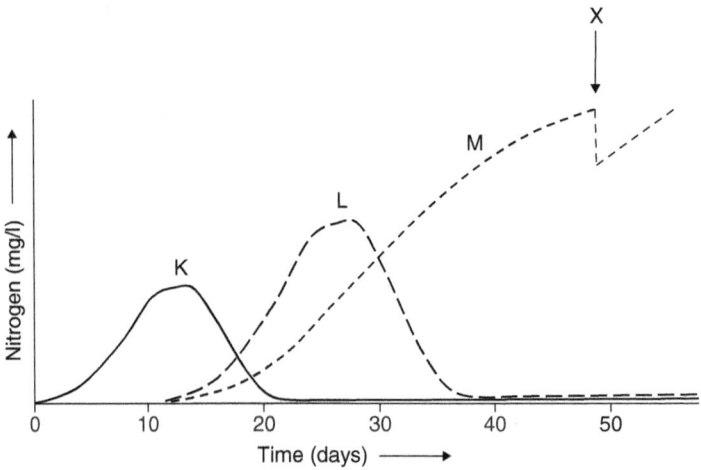

**Fig. 3.4.**

Table 3.3

| Graph line | A | B | C | D |
|------------|--------|---------|---------|---------|
| K | Nitrite | Ammonia | Nitrate | Ammonia |
| L | Nitrate | Nitrite | Nitrite | Nitrate |
| M | Ammonia | Nitrate | Ammonia | Nitrite |

    a.  A

    b.  B

    c.  C

    d.  D

**3.20a  In Fig. 3.4 what has occurred at point X?**

    a.  Some of the water has been changed

    b.  Fish have been added

    c.  The nitrifying bacteria have died

    d.  The change in M is a normal part of the process

# 4 Visitor Studies, Zoo Education and Zoo Research

This chapter contains questions on the behaviour of visitors in zoos, and the role of zoos in education and research.

## Foundation

**4.1f** **Biophilia is**

    a. the scientific name for a fear of animals

    b. a type of naturalistic enclosure design

    c. the scientific name of a group of rare amphibians

    d. the emotional affiliation of human beings to other living organisms

**4.2f** **Most people who visit a zoo or aquarium do so primarily**

    a. to learn about animals

    b. to support conservation

    c. to have a family day out and see some animals

    d. to feel a connection to nature

**4.3f** **Which of the following data for a zoo are most likely to reflect accurately the number of different individuals who visited over a particular period of time?**

    a. The number of admissions during a day

    b. The number of admissions during a week

© Paul A. Rees 2021. *Key Questions in Zoo and Aquarium Studies: A Study and Revision Guide* (P.A. Rees)
DOI: 10.1079/9781789249002.0004

c. The number of admissions during a month

d. The number of admissions during a year

**4.4f** **What effect do visitors have on the behaviour of animals in a zoo?**

a. A positive effect

b. A negative effect

c. No effect

d. It depends on the animal

**4.5f** **Which of the following does not assist visitor orientation in a zoo?**

a. Easily identifiable visitor meeting points

b. An effective public address system

c. Printed paper maps

d. Signage showing the layout of the zoo

**4.6f** **Research has shown that the most popular animals among zoo visitors are**

a. colourful amphibians

b. flightless birds

c. large mammals

d. poisonous reptiles

**4.7f** **Which of the following is unlikely to be the purpose of an interpretation board at a gorilla exhibit in a zoo?**

a. To explain the evolutionary relationships between gorillas and other primates

b. To describe the *in-situ* projects the zoo funds to help conserve gorillas in Africa

c. To list the food plants consumed by gorillas in the wild

d. To partially obscure the view of gorillas in the enclosure to reduce disturbance by visitors

4.8f  Signposts in zoos often contain pictures or silhouettes of animals instead of their names (Fig. 4.1). This is particularly useful for visitors who

a.  are lost

b.  are too young to read animal names

c.  do not understand the language in which the sign is written

d.  cannot read the animal names because of their age or understanding of the language

Fig. 4.1.

4.9f  The images of animals shown on the sign in Fig. 4.1 are collectively called

a.  pictograms

b.  ideograms

c.  hieroglyphs

d.  cyphers

**4.10f** A researcher examining the pathways visitors take when exploring a zoo is studying

   a. visitor infiltration

   b. visitor penetration

   c. visitor circulation

   d. visitor rotation

**4.11f** Which of the following does not publish peer-reviewed scientific research concerned with animals living in zoos?

   a. *International Zoo News*

   b. *Animal Welfare*

   c. *Journal of Applied Animal Welfare Science*

   d. *Applied Animal Behaviour Science*

**4.12f** A systematic analysis of ten years of zoo-themed research (Rose *et al.*, 2019) found that most of the subjects of this research were

   a. birds

   b. mammals

   c. reptiles or amphibians

   d. fishes or marine invertebrates

**4.13f** Figure 4.2 shows two identical enclosures at the same zoo, each of which contained a population of monkey species K of identical composition with respect to age structure, sex ratio and the total number of individuals. A scientist compared the activity budgets of the animals in Enclosure A with those of the animals in Enclosure B to determine whether or not the replacement of food dishes with food dispensers in Enclosure B had any effect. In this study the animals in Enclosure A are the

   a. experimental group

   b. control group

   c. replicate group

   d. duplicate group

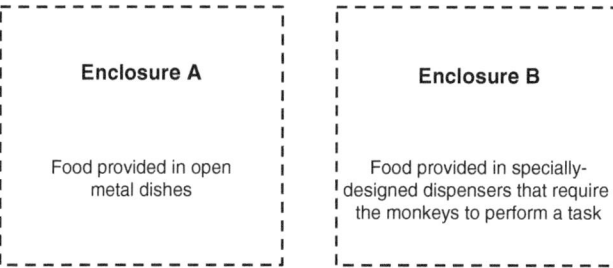

**Fig. 4.2.**

**4.14f** **A useful source of historical data relating to zoos, such as visitor numbers and the number of species kept, is**

a. *The International Zoo Yearbook*

b. *International Zoo News*

c. *Zoo Biology*

d. *Oryx*

**4.15f** **A researcher making a study of the educational value of a new penguin exhibit used a stopwatch to measure the amount of time each visitor spent in the exhibit watching the birds and compared her data with a similar study conducted on visitors to an older penguin exhibit in another zoo. This researcher was measuring the**

a. investment time

b. visitation time

c. dwell time

d. residence time

**4.16f** **The average amount of time visitors spend at an exhibit tends to increase if**

a. the animals are active

b. the animals are undergoing training

c. the exhibit allows public participation

d. any of the above is true

**4.17f** The weights of 10 fully-grown adult female tree pangolins (*Phataginus tricuspis*) were measured as part of a study of their growth and development in 25 zoos. The mean weight was 3.1kg with a standard deviation of 0.46kg. These two values are, respectively, measures of

a. dispersion and central tendency

b. central tendency and error

c. central tendency and dispersion

d. variation and central tendency

**4.18f** The Regional Studbook of the Association of Zoos and Aquariums (AZA) for the Central American spider monkey (*Ateles geoffroyi*) for 2010 recorded 211 animals living in 47 institutions: 73 males and 138 females. Which of the following statistical tests would you use to determine whether or not these values differ from a 1:1 ratio of males: females?

a. Dependent *t*-test

b. Chi-squared test

c. Independent *t*-test

d. Regression analysis

**4.19f** The sign in Fig. 4.3 is best described as

a. an infographic

b. a logogram

c. a phonogram

d. an informatic

**Fig. 4.3.**

**4.20f   The 'visitor attraction' model (Godinez and Fernandez, 2019) assumes that**

a.  animals are attracted to visitors when they appear at their enclosure

b.  visitors are attracted to active animals

c.  the presence of visitors reduces animal activity

d.  the presence of visitors increases animal activity

# Intermediate

4.1i    To access a new chimpanzee exhibit visitors walk along a narrow path lined on both sides with thick vegetation (simulating a forest) and then reach a gap in the vegetation from which they are able to see into part of the enclosure. This is known as

a.  a viewing point

b.  an access point

c.  an encounter point

d.  a glade

4.2i    Match the journals in Table 4.1 with the organisations with which they are most closely affiliated.

a.  A

b.  B

c.  C

d.  D

Table 4.1

| Organisation | A | B | C | D |
|---|---|---|---|---|
| Universities Federation for Animal Welfare | Applied Animal Behaviour Science | Animal Welfare | Journal of Zoo and Wildlife Medicine | Animal Welfare |
| European Association of Zoos and Aquaria | Journal of Zoo and Aquarium Research | Applied Animal Behaviour Science | Journal of Zoo and Aquarium Research | Journal of Zoo and Aquarium Research |
| International Society for Applied Ethology | Animal Welfare | Journal of Zoo and Aquarium Research | Applied Animal Behaviour Science | Applied Animal Behaviour Science |
| American Association of Zoo Veterinarians | Journal of Zoo and Wildlife Medicine | Journal of Zoo and Wildlife Medicine | Animal Welfare | Journal of Zoo and Wildlife Medicine |

**4.3i** Which of the following is not a good design principle for an interpretation board in a zoo?

a. Identify the target audience

b. Provide as much information as possible

c. Follow the zoo's brand guidelines

a. Locate the board in a prominent place

**4.4i** A study of the changes in the social behaviour of a group of lions (*Panthera leo*) living in a zoo, conducted over three years, is a

a. cross-sectional study

b. transverse study

c. latitudinal study

d. longitudinal study

**4.5i** Which of the following is least likely to aid zoo visitors during wayfinding?

a. A zoo map

b. A zoo guidebook

c. An interpretation board

d. A signpost

**4.6i** Use the data in Table 4.2 to calculate the average amount of time spent by Mr and Mrs Jones and their family looking at each of the exhibits in Metro Zoo and select the correct value from the list below (A-D). Assume that only the activities listed occurred, no time was used to travel between exhibits, and that the family visited all of the exhibits.

**Table 4.2**

| Arrival time | 10.15am |
|---|---|
| Departure time | 4.25pm |
| Time spent eating lunch | 45 minutes |
| Time spend on playground | 30 minutes |
| Number of exhibits | 76 |

a. 4.9 minutes

b. 4.5 minutes

c. 4.3 minutes

d. 3.9 minutes

**4.7i** **A display of confiscated felid skins, turtle shells, ivory products and elephant-hair bracelets may be used in a zoo to illustrate illegal wildlife trade regulated by**

a. CBD

b. CMS

c. CITES

d. ICRW

**4.8i** **Zoos located in the Member States of the European Union to which the Zoos Directive applies**

a. must provide educational courses for schools

b. must have an education function

c. must employ qualified teachers

d. must have an education centre

**4.9i** **The purpose of charging more for admission to a zoo on some days of the week and at some times of the year than others is to**

i. maximise income

ii. even out visitor attendance

iii. even out staffing requirements

iv. improve the visitors' experience by reducing crowding

a. i and ii

b. ii and iii

c. ii, iii and iv

d. achieve all of these things

**4.10i** **The normal flow of visitors along the main paths in Metro Zoo is shown in Fig. 4.4 along with the locations of the main exhibits. The times of the 15-minute keeper talks are shown in Table 4.3. Most visitors arrive at the zoo between 10.00 and 10.30am. You are a researcher examining the flow of visitors through the zoo and the effectiveness of the keeper talks. Which of the following recommendations would you make to the zoo managers to improve the visitor experience?**

a. Increase the length of all the keeper talks to 25 minutes

b. Set the times of the keeper talks to match the sequence in which most visitors encounter the exhibits

c. Give the Asiatic lions keeper talk first

d. Give the hunting dogs keeper talk last

**Table 4.3**

| Time | Location |
|------|----------|
| 11.00 | Hunting dog exhibit |
| 11.30 | Chimpanzee exhibit |
| 12.00 | Asiatic lion exhibit |
| 12.30 | Lemur exhibit |

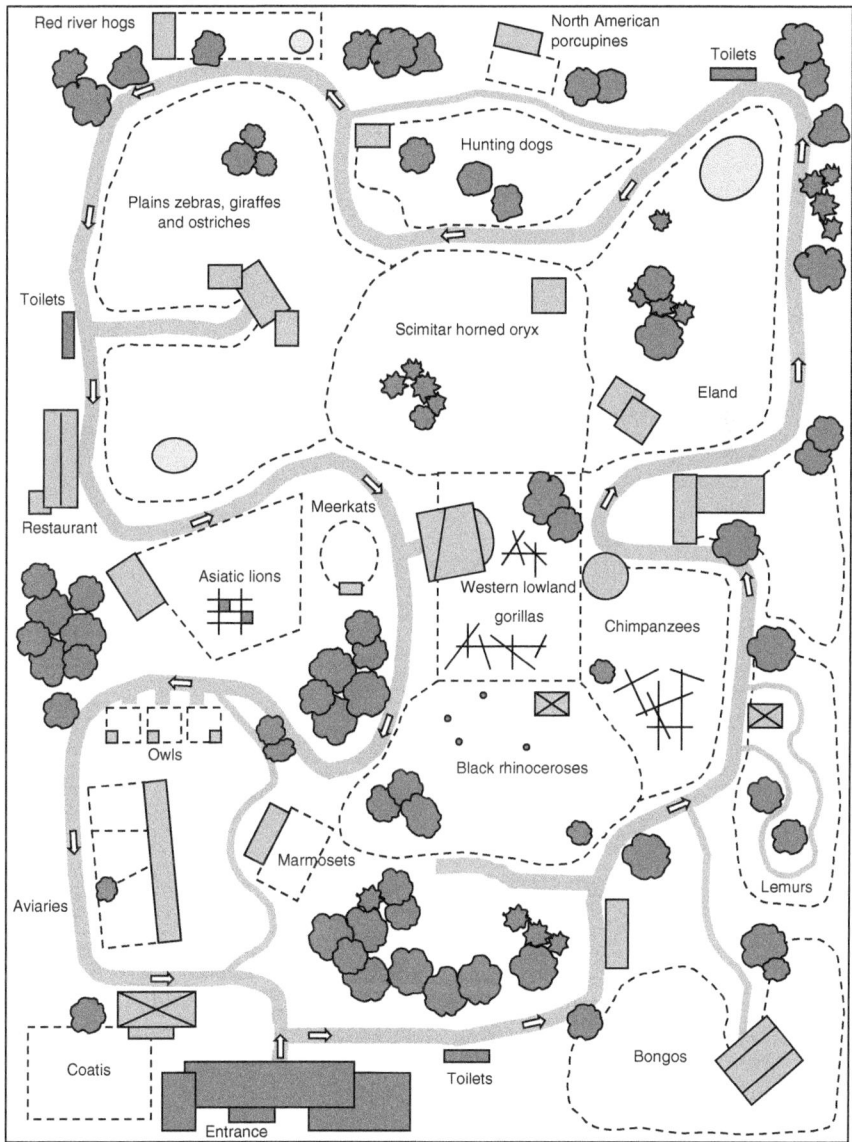

Fig. 4.4.

**4.11i   Red-necked wallabies (*Macropus rufogriseus*) and nilgai (*Boselaphus tragocamelus*) should not be kept in the same drive-through enclosure in a safari park because**

a.   the two species are from different zoogeographical regions and seeing them together sends a confusing educational message

b.   they may compete for food

71

  c. the presence of nilgai will make the wallabies less visible to visitors

  d. the two species are behaviourally incompatible

**4.12i** **The *International Zoo Yearbook* is published by**

  a. BIAZA

  b. WAZA

  c. EAZA

  d. the Zoological Society of London

**4.13i** **A great deal of information about the functioning of zoos, captive-breeding, animal husbandry and related matters can be found in reports, newsletters and other documents that are not widely available. This material is collectively termed the**

  a. black literature

  b. blue literature

  c. grey literature

  d. green literature

**4.14i** **Bloomsmith *et al.* (2006) studied the behavioural development of chimpanzees (*Pan troglodytes*) by making recordings of observations of 37 young individuals aged between 5 and 18 years living in 20 zoos. This was a**

  a. controlled study

  b. cross-sectional study

  c. meta-analysis

  d. longitudinal study

**4.15i** **In a study of visitor behaviour at North Carolina Zoo, Bitgood (1988) found that visitors were reluctant to walk down a path to an exhibit that took them away from the main path. He called this phenomenon**

  a. minor path reluctance

  b. principal route dominance

  c. major path adherence

  d. dominant path security

**4.16i** A researcher produced a graph showing the changes in the amount of time red river hogs (*Potamochoerus porcus*) spent feeding in a zoo enclosure using the axes in Fig. 4.5. Which of the following terms describe the variable on the *y*-axis (time spent feeding)?

    a. Discontinuous and dependent

    b. Discontinuous and independent

    c. Continuous and dependent

    d. Continuous and independent

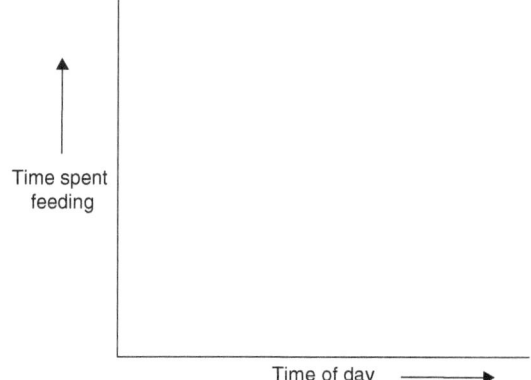

**Fig. 4.5.**

**4.17i** Three students were used to collect data for a study of the activity budgets of giraffes (*Giraffa camelopardalis*) in a zoo. On any particular day only one student made recordings. At the beginning of the study all three students independently collected the same data at the same time for a single day. The purpose of this was to ensure

    a. inter-observer reliability

    b. contra-observer reliability

    c. infra-observer reliability

    d. intra-observer reliability

**4.18i** **If a zoo reports an annual visitor attendance of 1,238,109 this is most likely to refer to**

a. the total number of different visitors to the zoo

b. the total number of visits made to the zoo including individuals that visited on more than one occasion during the year

c. the total number of different adult visitors to the zoo

d. the total number of families that visited the zoo

**4.19i** **Moss and Esson (2010) measured the relative interest that visitors showed in 40 species of animals kept at Chester Zoo, United Kingdom. Which of the following statements is least likely to be in agreement with their conclusions?**

a. Visitors found mammals more interesting than any other group

b. Visitors spent more time watching larger species than smaller species

c. Where exhibits contained flagship and non-flagship species there was no difference in the time visitors spent watching the animals that fell into these categories

d. Visitors spent more time watching active species than those that were inactive

**4.20i** **In a study of aquarium visitors, Zwinkels *et al.* (2009) found that visitors were more likely to spend time at an exhibit in the morning than in the afternoon. Some of this effect may be attributed to the fact that visitors' interest tends to decline as a visit progresses. In the science of visitor studies this phenomenon is known as**

a. afternoon apathy

b. museum fatigue

c. temporal disinterest

d. exhibit weariness

# Advanced

**4.1a** **Some zoos change their admission fees from day to day based on historical visitor data, weather, and other data to optimise the price point on any given day. This is known as**

a. seasonal pricing

b. active pricing

c. premium pricing

d. dynamic pricing

**4.2a** **Wells (2005) studied the effects of visitors on the behaviour of six zoo-housed gorillas (*Gorilla* sp.) during periods of low and high visitor density. Which of the following behaviours would you expect to increase with high visitor density?**

i. Time spent relaxing and resting

ii. Intragroup aggression

iii. Stereotypic behaviours

iv. Autogrooming

a. i, ii and iii

b. ii, iii and iv

c. i, iii and iv

d. i, ii and iv

**4.3a** **Bitgood (2006) has argued that a zoo visitor determines the value of an experience within the zoo by unconsciously calculating the ratio between the benefits and the costs. For example, a visitor incurs a cost walking to an isolated exhibit and receives a reward if the animals are visible on arrival. This is known as the**

a. unconscious cost principle

b. benefit principle

c. general value principle

d. profit and loss principle

**4.4a** **Which of the following statements about visitor behaviour in zoos in false?**

    a. Visitors tend to stay on the main paths

    b. Visitors are reluctant to backtrack to return to areas of the zoo they have already passed

    c. Visitors tend to move along one side of a path and are reluctant to move from one side to another

    d. Visitors tend to wander in different directions rather than walk in a straight line

**4.5a** **Some of the studies of animals living in zoos that have been published in the peer-reviewed literature have involved small numbers of animals. What was the smallest sample size?**

    a. One

    b. Two

    c. Three

    d. Four

**4.6a** **When visitors to the Arizona-Sonora Desert Museum were asked about how they found their way around the zoo (Shettel-Neuber and O'Reilly, 1981) almost two thirds said they**

    a. used the direction signs

    b. used a hand-held map of the zoo

    c. found their way by exploring

    d. used their knowledge of the zoo gained in previous visits

**4.7a** **An awareness of the themes and the organisation of the subject matter of a zoo is known as**

    a. conceptual orientation

    b. taxonomic orientation

    c. conceptual wayfinding

    d. taxonomic wayfinding

**4.8a** **The tendency for visitors to walk through an area containing exhibits in a museum, aquarium or similar location using as**

few steps as possible and take the shortest route from the entrance to the exit is known as

a. Melton's exit gradient theory

b. Bitgood's exit gradient theory

c. Melton's exit attraction theory

d. Bitgood's exit attraction theory

4.9a    Match the types of learning that may occur in zoo visitors with the examples of learning in Table 4.4.

Table 4.4

| During or after a zoo visit... | A | B | C | D |
|---|---|---|---|---|
| Priya learned facts about gorilla conservation in the wild | Behavioural | Cognitive | Affective | Cognitive |
| Kevin changed his attitude towards animal welfare | Cognitive | Behavioural | Behavioural | Affective |
| Katy was more likely to donate money to a conservation charity | Affective | Affective | Cognitive | Behavioural |

a. A

b. B

c. C

d. D

4.10a  The results of a study of visitor behaviour at a new aquarium exhibit for reef manta rays (*Mobula alfredi*) would be suitable for publication in the

a. *Zoo Educators Journal*

b. *International Zoological Education Journal*

  c. *International Zoo Educators Journal*

  d. *International Zoo Education Journal*

**4.11a** **The academic journal *Zoo Biology* was first published in**

  a. 1962

  b. 1972

  c. 1982

  d. 1992

**4.12a** **In a study of the social relationships within a group of three male cheetahs (*Acinonyx jubatus*) in a zoo Chadwick *et al.* (2013) calculated the probability of two cheetahs associating together by chance. We did this by using a computer simulation to plot randomly the positions of two imaginary animals within an imaginary enclosure and calculate the distance between them (Fig. 4.6). If the 'animals' were below a threshold distance apart they were considered to be associating. In Fig. 4.6 only the animals (dots) at position C are close enough to be considered to be associating based on the distance criterion illustrated. This process was repeated a large number of times so that the probability of two individuals associating by chance could be calculated. This type of simulation is known as a**

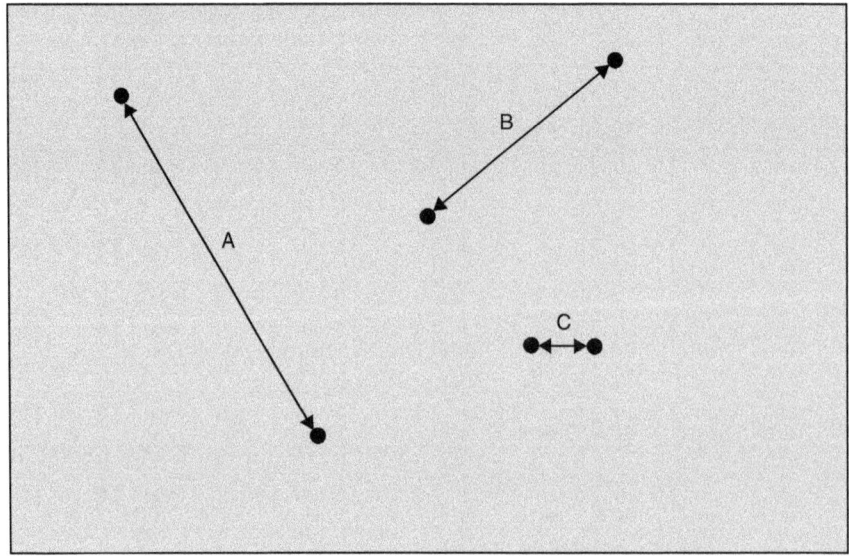

Fig. 4.6.

a. Monte Carlo simulation

b. Montenegro simulation

c. Montevideo simulation

d. Monte Cristo simulation

**4.13a** **A study of the relationship between the maximum daily environmental temperature and the frequency of stereotypic behaviour exhibited by an adult female Asian elephant (*Elephas maximus*) produced a Pearson product-moment correlation coefficient (*r*) of -0.326. Assuming this value is statistically significant it indicates that**

a. as temperature increases the frequency of stereotypic behaviour increases

b. as temperature decreases the frequency of stereotypic behaviour increases

c. the frequency of stereotypic behaviour is not affected by temperature

d. exposing elephants to lower temperatures may improve their welfare

**4.14a** **The result of a one-tailed test of the significance of the value of *r* in Q4.13a was *P*<0.05 and suggests that the relationship detected between the two variables**

a. was statistically significant and could have occurred by chance on fewer than 1 in 100 occasions

b. was statistically significant and could have occurred by chance on fewer than 5 in 100 occasions

c. was not statistically significant and could have occurred by chance on more than 95 in 100 occasions

d. was unclear and its statistical significance could not be determined

4.15a  A study that draws together and analyses data from a large number of studies that measured the effect of environmental enrichment on stereotypic behaviour in primates is a

a. mega-analysis

b. multi-analysis

c. meso-analysis

d. meta-analysis

4.16a  A scientist studied stereotypic behaviour in polar bears (*Ursus martimus*) living in three zoos (A, B and C). She used instantaneous scan sampling every 5 minutes to calculate the percentage of time spent stereotyping by each bear between 10.00am and 5.00pm each day (Table 4.5). The mean frequency of stereotypic behaviour was calculated from the 60 samples. However, the samples recorded were not independent so this is an example of

a. replication

b. duplication

c. pseudoreplication

d. repetition

Table 4.5

| | Zoo | | |
| --- | --- | --- | --- |
| | A | B | C |
| Number of bears | 2 | 1 | 3 |
| Observation days | 10 | 10 | 10 |
| Total days sampled per zoo | 20 | 10 | 30 |
| Total sample size | | 60 | |

4.17a  The data in Fig. 4.7 shows the relationship between maximum daily temperature and the frequency of dusting behaviour in Asian elephants (*Elephas maximus*) in a zoo. Which of the following statements about these data is false?

a.  At temperatures below 4°C dusting behaviour was not recorded

b.  There is a positive correlation between temperature and dusting frequency

c. Elephants dust more on warmer days to protect their skin from the sun

d. It is not possible to establish a causal relationship between dusting frequency and temperature from these data alone

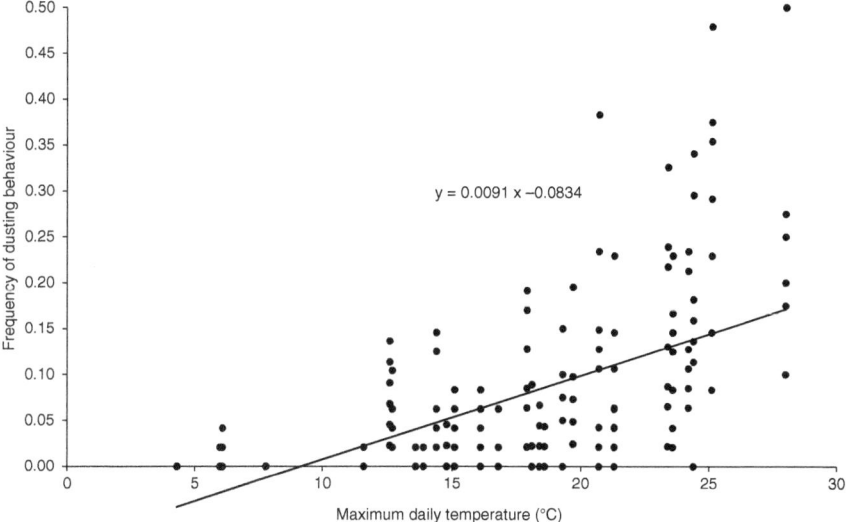

Fig. 4.7.

**4.18a A scientist recorded the amount of time a monkey spent exhibiting abnormal behaviours each day. During a pilot study lasting 7 days three different abnormal behaviours were observed and included in the definition of 'abnormal behaviour' in the ethogram. The study lasted for 6 months. In month 3 the scientist unconsciously included a fourth behaviour in the definition of 'abnormal' and in month 5 a fifth was included. These two additional behaviours were subtle changes that had probably been present in the animal's behavioural repertoire from the outset of the study but had not been noticed. As the observer used only one category for abnormal behaviour the data collected in months 3 and 4 were not comparable with those collected in the first two months and the data collected during both of these periods were not comparable with those collected from month 5 onwards. This is an example of**

a. observer drift

b. observer inconsistency

    c. standard error

    d. recorder error

4.19a A scientist measured the amount of time a black rhinoceros (*Diceros bicornis*) spent feeding using 5 minute scan samples and compared his results with data collected continuously using a video camera over the same period. This was done on eight days and the results are shown in Table 4.6.

Table 4.6

| Day | Percentage of time spent feeding | |
|-----|-------------|---------------|
|     | Scan sample | Video recording |
| 1 | 40 | 52 |
| 2 | 35 | 44 |
| 3 | 44 | 54 |
| 4 | 37 | 46 |
| 5 | 39 | 51 |
| 6 | 43 | 54 |
| 7 | 37 | 48 |
| 8 | 46 | 55 |

Assuming that the data collected from the video recording are accurate, use the options in Table 4.7 to indicate how the results from the scan sampling method can best be described.

Table 4.7

| | | Reliability | |
|---|---|---|---|
| | | High | Low |
| Validity | High | A | B |
| | Low | C | D |

    a. A

    b. B

    c. C

    d. D

**4.20a   The correlation coefficient between mean daily temperature and visitor numbers at a zoo was 0.87.**

    a.  This is statistically significant at the 5% level

    b.  This is statistically significant at the 1% level

    c.  This is not statistically significant

    d.  Considering this value alone, it is not possible to say whether or not this is statistically significant

# 5 Nutrition and Food Presentation

This chapter contains questions on the food provided for animals in zoos and aquariums, its nutritional value and presentation, and the anatomy and physiology of digestive systems.

## Foundation

**5.1f** **The correct sequence with which food passes through the stomach of a ruminant is**

a. oesophagus → abomasum → omasum → regurgitation → chewing cud → oesophagus → reticulum → rumen

b. oesophagus → rumen → reticulum → regurgitation → chewing cud → oesophagus → omasum → abomasum

c. oesophagus → rumen → omasum → regurgitation → chewing cud → oesophagus → reticulum → abomasum

d. oesophagus → reticulum → rumen → regurgitation → chewing cud → oesophagus → omasum → abomasum

**5.2f** **Which of the following is not a ruminant?**

a. Giraffe (*Giraffa camelopardalis*)

b. Moose (*Alces alces*)

c. Asian elephant (*Elephas maximus*)

d. Cape buffalo (*Syncerus caffer*)

© Paul A. Rees 2021. *Key Questions in Zoo and Aquarium Studies: A Study and Revision Guide* (P.A. Rees)
DOI: 10.1079/9781789249002.0005

**5.3f**   Much of a carnivore's requirement for which vitamin may be satisfied by eating the liver of its prey?

    a.  K

    b.  D

    c.  A

    d.  C

**5.4f**   *Marvin* was not supplied with logs or coarse feed and as a consequence his overgrown teeth began to interfere with his ability to feed. *Marvin* is not likely to be a

    a.  beaver (*Castor fiber*)

    b.  capybara (*Hydrochoerus hydrochaeris*)

    c.  crested porcupine (*Hystrix cristata*)

    d.  honey badger (*Mellivora capensis*)

**5.5f**   The object in Fig. 5.1 is located in the indoor accommodation of a group of giraffes. Its purpose is to supply the animals with

Fig. 5.1.

a. carbohydrates

b. minerals

c. protein

d. fats

**5.6f   Orchard-grown fruit differs from wild fruit in having**

a. a higher sugar content

b. a lower sugar content

c. a higher sugar content and a lower fibre and protein content

d. a lower sugar content and a higher fibre content

**5.7f   A feeding technique that gives animals a wide choice of foods is called**

a. *ad hoc* feeding

b. cafeteria-style feeding

c. complete feed-style feeding

d. *ad libitum* feeding

**5.8f   A piscivore predominantly eats**

a. fishes

b. insects

c. ants

d. leaves

**5.9f   Malnutrition may be the result of**

a. nutrient deficiencies in the diet

b. nutrient excesses in the diet

c. malabsorption of nutrients

d. any of the above

**5.10f   Which of the following is a micronutrient?**

a. Protein

b. Glucose

c. Iron

d. Saturated fat

**5.11f  Birds known as 'hardbills' feed primarily on**

a. seeds

b. insects

c. nectar

d. fruit

**5.12f  What types of animals sometimes feed their offspring crop milk?**

a. Some reptiles

b. Some mammals

c. Some birds

d. Some amphibians

**5.13f  The diets of farm animal species may be unsuitable for similar exotic species kept in a zoo because farm animal feeds have been developed to bring about**

a. rapid and efficient weight gain

b. high milk yield (in mammals)

c. high egg production (in birds)

d. all of the above

**5.14f  Marine mammals kept in captivity obtain most of their water from**

i.   the salt water in which they are kept

ii.  the drinking water provided

iii. their food

iv.  the metabolic breakdown of food

a. i

b. ii

c. iii

d. iii and iv

**5.15f** Junge et al. (2000) studied three juvenile chimpanzees (*Pan troglodytes*) that were raised indoors under skylights and consumed only breast milk. As a result they developed rickets due to a deficiency of vitamin

    a. $B_{12}$

    b. D

    c. C

    d. K

**5.16f** An animal described as a 'folivore' feeds on

    a. leaves

    b. bark

    c. flowers

    d. roots

**5.17f** San Diego Zoo grows plants to provide browse for its animals. Match the animals with the appropriate food plants in Table 5.1.

Table 5.1

|  | A | B | C | D |
|---|---|---|---|---|
| Koalas | Acacia | Eucalyptus | Eucalyptus | Bamboo |
| Giraffes | Bamboo | Bamboo | Acacia | Acacia |
| Red pandas | Eucalyptus | Acacia | Bamboo | Eucalyptus |

    a. A

    b. B

    c. C

    d. D

**5.18f** The form of chemoreception usually called taste is also known as

    a. gustation

    b. gestation

    c. appetence

    d. olfaction

**5.19f** **An amino acid that must be present in the diet because it cannot be synthesised by an animal's body is called**

a. an important amino acid

b. a critical amino acid

c. an essential amino acid

d. a fundamental amino acid

**5.20f** **Keepers sometimes cut up fruit into small pieces and spread these around an enclosure by throwing them in different directions so that the animals must search for them. This is known as**

a. *ad libitum* feeding

b. dispersion feeding

c. scatter feeding

d. *ad hoc* feeding

# Intermediate

**5.1i** **In herbivores, gastroliths, enteroliths and uroliths may result from overfeeding with calcium and are**

a. lesions

b. abscesses

c. tumours

d. concretions

**5.2i** **Vitamin B$_1$ deficiency is rare in animals but may occur in pinnipeds because some fishes contain enzymes that destroy this vitamin. These enzymes are known as**

a. kinases

b. thiaminases

c. lipases

d. proteases

**5.3i**   Which of the following feeding regimes is most appropriate for captive lions (*Panthera leo*)?

a.  Gorge and fast feeding

b.  Daily nocturnal feeding

c.  Daily diurnal feeding

d.  Cafeteria-style feeding

**5.4i**   The consumption of non-food items is a disorder known as

a.  euphagia

b.  tikka

c.  dysphagia

d.  pica

**5.5i**   The practice of rearing insects on a nutrient-rich diet to improve their nutritional value – because their guts retain some of the nutrients – before feeding them to amphibians is known as

a.  gut loading

b.  nutrient enrichment

c.  nutrient up-scaling

d.  nutrient loading

**5.6i**   The unapproved gradual alteration of an animal's diet in a zoo, possibly due to a lack of communication between nutritionists and those that feed the animals, is called

a.  dietary shift

b.  dietary drift

c.  dietary change

d.  dietary transition

**5.7i**   Which of the following may suppress reproduction in mammals?

a.  Overfeeding only

b.  Underfeeding only

c.  Overfeeding or underfeeding

d.  Nutritional factors do not suppress reproduction

**5.8i** A metabolomics assay measures

    a. the utilisation of energy in food

    b. the protein content of food

    c. the chemical fingerprints left by cellular processes

    d. the metabolic rate of an animal

**5.9i** The nutrient known as α-tocopherol is a form of vitamin

    a. E

    b. K

    c. C

    d. B

**5.10i** Which of the following are not brands of exotic animal foods?

    i. *Mazuri*

    ii. *Kasper*

    iii. *Nutrazu*

    iv. *OptiBird*

    v. *DK Zoological*

    vi. *Garvo*

    a. i and v

    b. ii and vi

    c. iii and vi

    d. All are brands of exotic animal foods

**5.11i** Dietary intake is calculated as

    a. weight of food provided

    b. *weight of food offered – weight of food not eaten*

    c. $\dfrac{weight\ of\ food\ offered}{weight\ of\ food\ not\ eaten} \times 100$

    d. $\dfrac{weight\ of\ food\ offered - weight\ of\ food\ not\ eaten}{weight\ of\ food\ offered} \times 100$

**5.12i** **Cafeteria-style feeding is discouraged for animals in zoos because**

a. it wastes food

b. given a choice animals rarely select a balanced diet

c. it may contain food items that the animals find unpalatable

d. it is expensive

**5.13i** **Hand-reared exotic ruminants may be fed colostrum from domestic cows so that they acquire**

a. proteins

b. immunosuppressants

c. essential amino acids

d. immunoglobulins

**5.14i** **The diet of an exotic species kept in a zoo is often based on the nutritional needs of a well-known domesticated animal referred to as a**

a. archetypal species

b. exemplar species

c. model species

d. prototype species

**5.15i** **Which of the following species is not a hindgut fermenter?**

a. European bison (*Bison bonasus*)

b. Black rhinoceros (*Diceros bicornis*)

c. Malaysian tapir (*Acrocodia indica*)

d. Grevy's zebra (*Equus grevyi*)

**5.16i** **Cole *et al.* (2020) reported that a giant anteater (*Myrmecophaga tridactyla*) (Fig. 5.2) fed on a commercial insectivore food had an excess of vitamin D$_3$ in its diet. This animal was suffering from**

a. hypovitaminosis

b. avitaminosis

c. hypervitaminosis

d. multivitaminosis

Fig. 5.2.

**5.17i A maintenance ration may be described as that which will keep an animal that**

a. is active and in good health in the same condition and at the same weight for a week

b. is sleeping and in good health in the same condition and at the same weight for an indefinite period

c. is in a resting condition and in good health in the same condition and at the same weight for an indefinite period

d. is in a resting condition and in good health in the same condition and at the same weight for a week

**5.18i The minimum rate of energy usage compatible with life is called the**

a. standard metabolic rate

b. resting metabolic rate

c. fasting metabolic rate

d. basal metabolic rate

**5.19i** **Which of the following statements about food provisioning for zoo animals is false?**

a. The palatability of food is largely unaffected by storage conditions

b. It is difficult to control the diets of different species provided with food in a multi-species exhibit where they share the same enclosure

c. Frozen fish should not be thawed by washing as this may result in the leaching out of water-soluble nutrients

d. Vitamins may be lost from food stored for a long period

**5.20i** **Which of the following elements is most important in egg shell production in birds?**

a. Magnesium

b. Potassium

c. Calcium

d. Phosphorus

## Advanced

**5.1a** **A lizard may refuse to eat**

i. just before shedding its skin

ii. before egg laying

iii. when the enclosure is too cold

iv. during the mating season

v. when experiencing stress due to over-handling

a. i, ii, and iii

b. ii, iii and v

c. i, ii, iii and v

d. i, ii, iii, iv and v

**5.2a** **Which of the following is not a component of dietary fibre?**

a. Cellulose

b. Chitin

    c. Pectin

    d. Lignin

**5.3a** **Which of the following were benefits of restricting calorific intake in rhesus macaques (*Macaca mulatta*) according to a study by Colman *et al*. (2009)?**

    i. Reduced incidence of diabetes

    ii. Reduced incidence of cancer

    iii. Reduced incidence of cardiovascular disease

    iv. Reduced incidence of brain atrophy

    v. Increased life expectancy

    a. i, iii and v

    b. i, ii, iv and v

    c. i, ii and iii

    d. i, ii, iii, iv and v

**5.4a** **The black rhinoceros (*Diceros bicornis*) (Fig. 5.3) is susceptible to haemosiderosis so should be fed a diet containing a low level of**

Fig. 5.3.

a. iron

b. copper

c. magnesium

d. calcium

**5.5a** **Which of the following statements about the dietary requirements of animals is false?**

a. Dietary requirements may change if an animal is sick

b. A female mammal may need more and different foods when pregnant or suckling

c. In polymorphic species males often need more food than females

d. Homiotherms need less energy than poikilotherms of the same size

**5.6a** **Which of the following statements about faecal condition scoring is false?**

a. It may use photographs

b. It may use numerical values

c. It includes the results of chemical analyses

d. Systems differ between taxa

**5.7a** **Pinnipeds maintained in fresh water commonly exhibit hyponatraemia or**

a. potassium deficiency

b. iron deficiency

c. sodium deficiency

d. magnesium deficiency

**5.8a** **What type of captive marine mammals can be fed on a diet consisting of lettuce, alfalfa, cabbage and aquatic plants such as water hyacinths?**

a. Sirenians

b. Cetaceans

c. Phocids

d. Odobenids

**5.9a** **Polar bears (*Ursus maritimus*)(Fig. 5.4) are considered to have an exceptionally high dietary requirement for vitamin**

a. A

b. $B_{12}$

c. C

d. D

Fig. 5.4.

**5.10a** **Animals kept in zoos may be at risk of poisoning by toxic plants. Experienced free-living wild animals are able to avoid poisoning by secondary plant compounds by using which of the following strategies?**

i. Avoidance

ii. Dilution

iii. Gastrointestinal degradation

iv. Detoxification

v. Anabolism

vi. Glycolysis

a. i, ii and iii

b. ii, iii, iv and vi

c. i, iv, v and vi

d. i, ii, iii and iv

**5.11a The acid-insoluble ash (AIA) marker technique is a method of determining**

a. the protein content of food

b. the dry matter digestibility of food

c. the energy content of browse

d. the acidity of food

**5.12a Which of the following may cause a vitamin deficiency?**

i. Nutritional inadequacy

ii. Malabsorption

iii. The effect of pharmacological agents

iv. Abnormalities of vitamin metabolism

v. Utilisation in metabolic pathways

a. i, ii and iv

b. i, iii and iv

c. i, ii, iii and iv

d. i, ii, iii, iv and v

**5.13a Hand-reared pinnipeds should be given milk that is**

a. low in lactose because they produce little lactase

b. high in lactose because they produce large quantities of lactase

c. low in protein because they produce little protease

d. low in fructose because they produce little fructase

**5.14a Which of the following computer packages are designed to facilitate the management of animal diets in zoos?**

a. *Zootrition* and *Fauna*

b. *Nutrition* and *Zootrition*

    c. *Fauna* and *Nutrition*

    d. *Diet* and *Zootrition*

**5.15a Mass-specific metabolic rate is measured as**

    a. joules/second

    b. $O_2$ cm³/minute

    c. joules/kg

    d. $O_2$ l/kg/hour

**5.16a Which of the following pairs of terms mean the same thing?**

    a. Hedyphagia and euphagia

    b. Nutritional wisdom and aphasia

    c. Nutritional wisdom and euphagia

    d. Dysphagia and hedyphagia

**5.17a In the formula**

$$X = \frac{A - B}{A} \times 100$$

**where,**

**A = dry weight of food eaten in one week**

**B = dry weight of faeces produced in one week**

**what is X?**

    a. Feeding efficiency

    b. Assimilation efficiency

    c. Ecological efficiency

    d. Absorption efficiency

**5.18a If an animal eats a novel food and is sick immediately afterwards it may become conditioned to the smell, sight or taste of this food and avoid consuming it in future as a result of**

    a. avertable conditioning

    b. avoidance imprinting

    c. aversive learning

    d. repellent conditioning

**5.19a Which of the following diseases is not caused by a nutritional deficiency?**

    a. Haemochromatosis

    b. Aspergillosis

    c. Metabolic bone disease

    d. Steatitis

**5.20a Preparing food by peeling and chopping it into small pieces (Fig. 5.5) can have a negative effect on animal nutrition because**

    i. it exposes a large surface area to the possibility of bacterial contamination

    ii. it may result in the loss of vitamins and other nutrients

    iii. it makes the food more susceptible to desiccation

    iv. it deprives the animals of the enrichment and learning opportunities provided by having to manipulate whole food items

    a. i, ii and iv

    b. i, iii and iv

    c. i, ii, and iii

    d. All of these are true

Fig. 5.5.

# 6 Reproductive Biology and Genetics

This chapter contains questions on the biology of reproduction, population dynamics and the genetics of animals living in zoos.

## Foundation

**6.1f** All of the genes in a particular population of a species are collectively called its

    a. gene pool

    b. gene pond

    c. gene bank

    d. gene reservoir

**6.2f** Which of the following statements about giant pandas (*Ailuropoda melanoleuca*) is false?

    a. They are monoestrous

    b. They exhibit delayed implantation

    c. They usually produce a single cub

    d. The cubs are born at an advanced stage of development

**6.3f** Which of the following is not an interspecies hybrid?

    a. Wholphin

    b. Zonkey

    c. Liger

    d. White tiger

© Paul A. Rees 2021. *Key Questions in Zoo and Aquarium Studies: A Study and Revision Guide* (P.A. Rees)
DOI: 10.1079/9781789249002.0006

**6.4f**  The cloaca is the terminal part of the gut where the alimentary canal, urinary and reproductive systems open into a common aperture in

a. amphibians and reptiles

b. birds, reptiles and amphibians

c. birds, reptiles, amphibians and monotremes

d. birds and reptiles

**6.5f**  When is the best time to inseminate a female mammal

a. 1 day before ovulation

b. 1 day after ovulation

c. The optimum time varies between species

d. 5 days after ovulation

**6.6f**  An ectopic pregnancy is one in which

a. a fertilised egg fails to implant

b. a fertilised egg implants somewhere other than in the uterus

c. more than one foetus is produced

d. the pregnancy is terminated by spontaneous abortion

**6.7f**  The potential capacity of an organism for reproduction is its

a. fecundity

b. intrinsic rate of natural increase

c. natality

d. birth rate

**6.8f**  In the past it was common to find some zoos keeping closely related species in the same cages and interspecific hybrids were sometimes produced as a result. A tigon is a hybrid formed from a mating between

a. a female tiger and a male lion

b. a tiger of either sex and a lion of the opposite sex

c. a lion x tiger hybrid and a pure bred tiger

d. a male tiger and a female lion

**6.9f    Ovulation in mammals is triggered by**

   a.  a surge in luteinising hormone

   b.  a surge in follicle-stimulating hormone

   c.  a fall in luteinising hormone

   d.  a fall in follicle-stimulating hormone

**6.10f   Which of the following might be administered to a pregnant large mammal to induce labour?**

   a.  Prolactin

   b.  Oestrogen

   c.  Oxytocin

   d.  Progesterone

**6.11f   Which of the following animals is not oviparous?**

   a.  An ostrich

   b.  An alligator

   c.  A dragonfly

   d.  A wallaby

**6.12f   A zoo population management strategy that allows animals to reproduce and keeps the group size constant by selectively removing individuals is called**

   a.  breed and kill

   b.  breed and cull

   c.  reproduce and kill

   d.  reproduce and cull

**6.13f   Which of the following statements about brood patches in birds is false?**

   a.  They contain a large number of blood vessels

   b.  They are used for incubation

   c.  They occur in females only

   d.  Most bird species have evolved brood patches

**6.14f   Testiconid mammals have**

   a.  exceptionally large testes for their body size

   b.  no testes as a result of a congenital abnormality

   c.  only one descended testicle

   d.  internal testicles

**6.15f   Flehmen is part of a behaviour that allows males of some mammal species to determine whether or not females are in oestrus using an organ in the nasal cavity to detect odours (Fig. 6.1). This organ is called the**

   a.  Stephenson's organ

   b.  Williamson's organ

   c.  Jacobson's organ

   d.  Davidson's organ

Fig. 6.1.

**6.16f** Match the definitions in Table 6.1 with the correct list of terms in columns A-D.

Table 6.1

| Definition | A | B | C | D |
|---|---|---|---|---|
| The individual's appearance and behaviour | Phenotype | Phenotype | Genotype | Genome |
| Genetic makeup of an individual in relation to a single gene or a small number of genes | Genotype | Genome | Phenotype | Phenotype |
| All of the genetic material of an organism | Genome | Genotype | Genome | Genotype |

    a. A

    b. B

    c. C

    d. D

**6.17f** Which of the following is not a secondary sexual characteristic in male fish?

    a. Bright coloration

    b. Gonads

    c. Increased size

    d. Accessory fins

**6.18f** Temperature-dependent sex determination does not occur in

    a. turtles

    b. alligators

    c. most birds

    d. frogs

**6.19f  Match the name of each animal in Table 6.2 with the correct term for its offspring**

Table 6.2

| Animal type | A | B | C | D |
|---|---|---|---|---|
| Llama | Calf | Cria | Kit | Cria |
| Koala | Joey | Joey | Cria | Kit |
| Beaver | Kit | Kit | Joey | Calf |
| Manatee | Cria | Calf | Calf | Joey |

    a.  A

    b.  B

    c.  C

    d.  D

**6.20f  Fossa (*Cryptoprocta ferox*) pups are born as miniature adults. This form of reproduction is known as**

    a.  oviparity

    b.  ovoviviparity

    c.  ovuliparity

    d.  viviparity

# Intermediate

**6.1i  A graph showing the pattern of deaths across different age groups within a cohort of animals of the same species (Fig. 6.2) is called**

    a.  a survivorship curve

    b.  a mortality curve

    c.  a death rate curve

    d.  a logistic curve

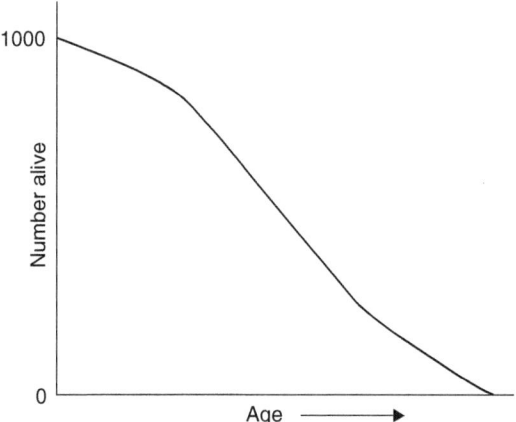

**Fig. 6.2.**

**6.2i** **The term 'autosome' refers to**

    a. an inclusion found in the cytoplasm of an egg cell

    b. the midsection of a spermatozoan (between the head and the tail)

    c. a chromosome that is not a sex chromosome

    d. a small fragment of DNA

**6.3i** **Complete the following definition using the appropriate term from the list below: 'The effective population size may be defined as the size of an ideal population that would lose genetic variation by .......... at the same rate'.**

    a. mutation

    b. genetic drift

    c. immigration

    d. emigration

**6.4i** **Calculate the effective population size ($N_e$) of the following population of animals using the formula and data below:**

$$N_e = \frac{4\left(N_f \times N_m\right)}{\left(N_f + N_m\right)}$$

$N_m$ (number of breeding males) = 98

$N_f$ (number of breeding females) = 76

a. 43

b. 171

c. 192

d. 163

**6.5i**   **A dynamic life table is constructed from**

a. data from a cohort of animals born at the same time followed throughout their lives until they die

b. the ages of all of the animals present in a population at a particular point in time

c. a survivorship curve

d. a static life table

**6.6i**   **The entire zoo population of a species of rare gazelle has been taken from the same locality in Africa and the individuals used to establish a captive breeding programme. When the genomes of the individuals were sequenced it was clear that the frequencies of some alleles in this zoo population were unusually high while those of other alleles were unusually low, compared with those in the remaining wild populations. This loss of genetic variation is known as**

a. the originator effect

b. the creator effect

c. the founder effect

d. the gene pool effect

**6.7i**   **In 2006 a Komodo dragon (*Varanus komodoensis*) at Chester Zoo produced fertile eggs without ever having mated with a male. This phenomenon is knows as**

a. parthenogenesis

b. hermaphroditism

c. ovoviviparity

d. oviparity

6.8i  **Which of the following statements about induced ovulation in mammals is false?**

   a. It may be caused by the physical act of copulation in some species

   b. It is not known to occur in camelids

   c. The physical presence of a conspecific male is required for it to occur in some species

   d. It may be caused by the presence of semen in some species

6.9i  **Which of the following is likely to be a sign of imminent parturition in a mammal?**

   a. Swollen or distended nipples

   b. Swollen and distended vulva

   c. Mammary secretions

   d. All of the above are signs of imminent parturition

6.10i  **Which of the following is a measure used to assess the genetic importance of an individual animal within a population by considering the number of relatives it has in that population and the degree of relatedness?**

   a. Kinship index

   b. Coefficient of inbreeding

   c. Mean kinship

   d. Coefficient of relatedness

6.11i  **The coefficient of relatedness between an individual animal and its clone is**

   a. zero

   b. 0.25

   c. 0.5

   d. 1.0

6.12i  **The age structure of a population of zoo animals that are part of a captive breeding programme is illustrated in Fig. 6.3. From the evidence in this graph, the population size appears to be**

   a. decreasing

   b. increasing slowly

c. increasing rapidly

d. stable

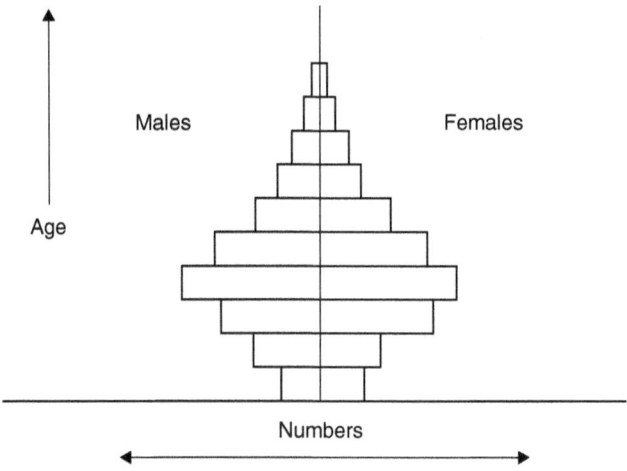

**Fig. 6.3.**

**6.13i  In a population in which all of the individuals are capable of reproduction, the effective population size is the same as the actual population size when**

a. the sex ratio is 1:1

b. there are twice as many males as females

c. there are twice as many females as males

d. none of the above is true

**6.14i  In relation to sex chromosomes, female birds are**

a. homozygotic

b. heterogametic

c. homogametic

d. heterozygotic

**6.15i  In the non-luteal phase of the oestrous cycle in the Asian elephant (*Elephas maximus*)**

a. there is one surge in the level of luteinising hormone and this induces ovulation

b. there are two surges in the level of luteinising hormone and the first induces ovulation

    c. there are two surges in the level of luteinising hormone and the second induces ovulation

    d. the level of luteinising hormone does not affect ovulation

**6.16i** **Which of the following statements about birds is false?**

    a. A female bird that is unable to expel an egg is said to be 'egg bound'

    b. Most female birds have two functional oviducts throughout their life

    c. The ovary increases in size during the breeding season

    d. The vitelline membrane surrounds the yolk in the egg

**6.17i** **The device in Fig. 6.4 is a PCR machine which is used**

    a. to amplify segments of genetic material

    b. to sort semen based on the sex chromosome carried

    c. to incubate the eggs of birds and reptiles

    d. in artificial insemination

Fig. 6.4.

**6.18i** **Which of the following is an acronym for software that analyses data relating to breeding populations of animals in zoos?**

    a. ARKPOP

    b. POPSTAT

    c. BREED

    d. SPARKS

**6.19i** **In a zoo population of 87 rare parrots only two individuals had an allele (k) that conferred resistance to a particular disease. Both of these birds died during breeding transfers. Allele k was lost from the population as a result of**

    a. gene flow

    b. gene mutation

    c. genetic drift

    d. gene extinction

**6.20i** **Match the description of the mating system in the first column of Table 6.3 with its name.**

Table 6.3

| Description | A | B | C | D |
|---|---|---|---|---|
| One female mates with several males in the same breeding season | Polyandry | Promiscuity | Polygyny | Polygynandry |
| Both sexes have several mating partners in the same breeding system | Polygynandry | Polyandry | Promiscuity | Promiscuity |
| One male mates with several females in the same breeding season | Polygyny | Polygynandry | Polyandry | Polyandry |
| Males and females mate randomly with several partners | Promiscuity | Polygyny | Polygynandry | Polygyny |

a. A

b. B

c. C

d. D

# Advanced

**6.1a** **Artificial selection for docility occurs in captive breeding programmes because more aggressive or stressed animals are**

a. more likely to be injured and die from trauma

b. more likely to reproduce poorly in confined spaces in captivity

c. harder to handle so animal managers might favour more docile individuals

d. All of the above are true

**6.2a** **As inbreeding increases**

a. genetic load decreases

b. genetic load increases

c. mutation rate increases

d. mutation rate decreases

**6.3a** **In the American alligator (*Alligator mississippiensis*) the thermosensor protein TRPV4 determines**

a. the clutch size

b. duration of sexual arousal

c. the timing of egg laying behaviour

d. the sex of offspring

**6.4a** **The smallest population size required to provide some specified probability that a population will survive for a given period of time is known as the**

a. lowest practicable population

b. minimum effective population

c. minimum viable population

d. smallest survivable population

6.5a The coefficient of genetic relatedness (r) is a measure of the extent to which two animals are genetically related: the probability that two individuals have inherited the same allele from a common ancestor. The relationships between seven individual animals are shown in Fig. 6.5. Individuals A, B, E and F are unrelated.

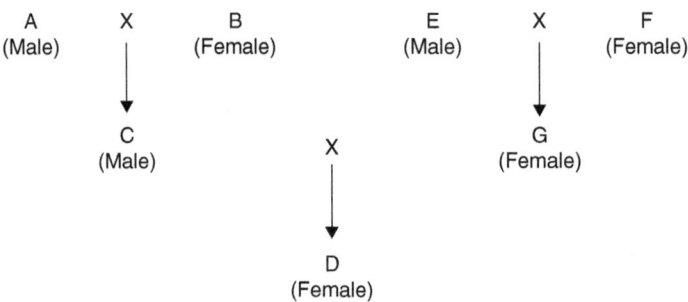

Fig. 6.5.

What is the value of r for C and E?

a. 0.5

b. 0.25

c. 0.125

d. 0

6.6a A group of brown capuchin monkeys (*Sapajus apella apella*) kept at Edinburgh Zoo, Scotland, is shown in Fig. 6.6. What is the value of r for the two capuchins *Anita* and *Lindo*?

a. 1.0

b. 0.5

c. 0.25

d. 0

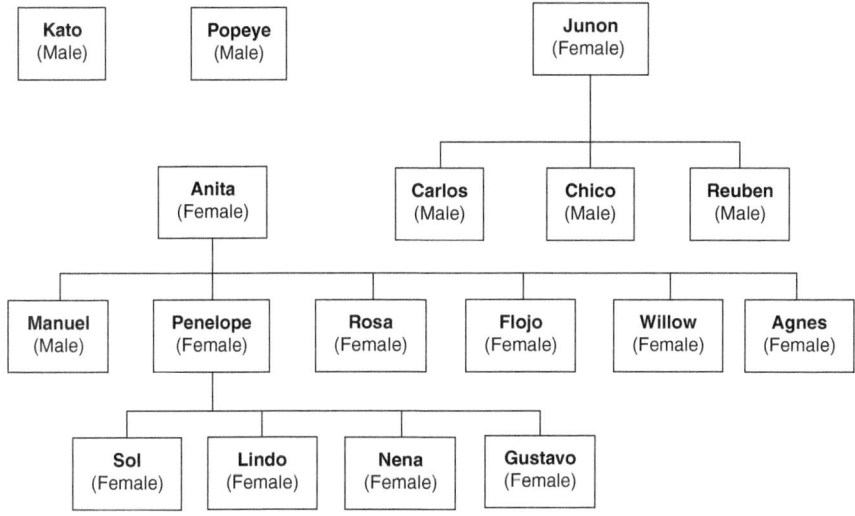

Fig. 6.6.

**6.7a** **If the coefficient of relatedness (*r*) between two animals is 0.25, these animals are**

a. half-siblings

b. uncle and niece

c. grandparent and grandchild

d. any of the above

**6.8a** **The effective population size can be increased by**

a. producing an equal number of offspring per female

b. managing for equal sex ratio for breeders in the population

c. rotating males among groups

d. all of the above

**6.9a** **In a life table, if there are 987 two-year old animals and 850 three-year old animals, the age-specific death rate of two-year old animals is**

a. 987

b. 987-850

c. 850

d. 137/987

**6.10a  Which of the following pieces of equipment would be most useful for sexing monomorphic species of birds?**

a.  Oscilloscope

b.  Laparoscope

c.  Microscope

d.  Stethoscope

**6.11a  Which of the following hormones causes milk production in mammals and where is it produced?**

a.  Oestrogen from the ovaries

b.  Prolactin from the pituitary gland

c.  Progesterone from the adrenal glands and ovaries

d.  Follicle-stimulating hormone from the pituitary gland

**6.12a  Harem species cause population problems for zoos because**

a.  they always produce more female offspring than male offspring

b.  they always produce more male offspring than female offspring

c.  the zoo population as a whole will contain surplus males because male and female offspring are produced in equal proportions

d.  the zoo population as a whole will contain surplus females because male and female offspring are produced in equal proportions

**6.13a  Which of the following species experiences a copulatory tie during mating?**

a.  Bush dog (*Speothos venaticus*)

b.  African elephant (*Loxodonta africana*)

c.  Western grey kangaroo (*Macropus fuliginosus*)

d.  Prevost's squirrel (*Callosciurus prevostii*)

**6.14a  Which of the following is least useful as a source of material for determining the sex of a bird?**

a.  A feather plucked from a live bird

b.  A feather collected from the cage floor

c.  A blood sample from the claw

d.  A blood sample from some other part of the body

**6.15a** A zoo has a herd of American bison (*Bison bison*) consisting of 12 adult males, 29 adult females, 6 juvenile males and 7 juvenile females. The ratio of males to females is best described as

    a. 12:29

    b. 2:1

    c. 1:2

    d. 18:36

**6.16a** Which of the following zoo staff is most likely to calculate inbreeding coefficients during the course of her work?

    a. A studbook keeper

    b. A zoo vet

    c. A zoo curator

    d. A head keeper

**6.17a** If a zoo keeps 1.2.3 fish of a particular species the '3' represents

    a. the total number of the species

    b. the number of males

    c. the number of females

    d. the number of unknown sex

**6.18a** A zoo held 7.4.2 scarlet ibises (*Eudocimus ruber*) on 1st January 2020. During the year the following changes occurred:

| | |
|---|---|
| Deaths | 0.1.0 |
| Births | 1.2.0 |
| Arrivals | 1.2.0 |
| Departures | 0.1.1 |

How many animals were present at the end of 2020?

    a. 5.4.1

    b. 9.5.1

    c. 9.6.3

    d. 9.6.1

6.19a  In brown bears (*Ursus arctos*) albinism is caused by a recessive autosomal allele. What is the probability that an albino female will produce an albino cub if she mates with a normal male whose mother was an albino?

    a.  0

    b.  0.25

    c.  0.50

    d.  0.75

6.20a  The chromosomes shown in Fig. 6.7 are the sex chromosomes of a male mammal. Males are heterogametic and possess two different sex chromosomes (XY). The labels r, s, t, u and v indicate the positions of recessive alleles for genes located on the X chromosome. Alleles on the Y chromosome are not shown. Which genes are sex-linked?

    a.  t, u and v

    b.  r and s

    c.  r only

    d.  r, s and t

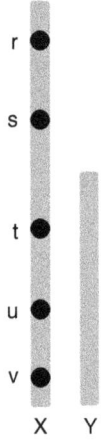

Fig. 6.7.

# 7 Conservation Breeding and Assisted Reproductive Technologies

This chapter contains questions on the role of zoos in conservation (captive) breeding programmes and the technologies that have been developed to assist in the reproduction of rare species.

## Foundation

**7.1f** The study of the effects of freezing on organisms – including the preservation of sperm and ova to help prevent the extinction of species – is called

    a. cryptozoology

    b. cryobiology

    c. thermobiology

    d. gerontology

**7.2f** *In-vitro* fertilisation results in the formation of an embryo in

    a. fresh water

    b. the uterus

    c. the laboratory

    d. the oviduct

**7.3f** Which of the following species has not been saved from extinction by captive breeding?

    a. Mauritius kestrel (*Falco punctatus*)

    b. California condor (*Gymnogyps californianus*)

© Paul A. Rees 2021. *Key Questions in Zoo and Aquarium Studies: A Study and Revision Guide* (P.A. Rees)
DOI: 10.1079/9781789249002.0007

    c. Przewalski's horse (*Equus ferus przewalskii*)

    d. Fish hawk (*Pandion haliaetus*)

**7.4f**    **Hornbills will nest in a wooden barrel accessed via a hole in its side. In addition they need access to a plentiful supply of mud to**

    a. completely seal up the hole with the female inside

    b. partially seal up the hole with the female inside

    c. partially seal up the hole with the male and female inside

    d. spread on the floor of the barrel

**7.5f**    **Complete the following sentence with one of the options listed below: '*Patula* is a genus of rare ........... that has been bred in captivity'.**

    a. African dung beetles

    b. European honey bees

    c. tropical land snails

    d. swallowtail butterflies

**7.6f**    **Which vertebrate was cloned in 1958 at the University of Oxford using intact nuclei from somatic cells?**

    a. A *Xenopus* tadpole

    b. A sheep

    c. A stickleback (*Gasterosteus*)

    d. An axolotl (a salamander: *Ambystoma*)

**7.7f**    **EEPs are the most intensive type of population management programmes for species kept in**

    a. EAZA zoos

    b. AZA zoos

    c. ARAZPA zoos

    d. ZAA zoos

**7.8f**    **The WWT specialises in breeding rare**

    a. parrots

    b. waterbirds

c. reptiles

d. bats

**7.9f    An ESU is an**

a. ecologically significant unit

b. environmentally significant unit

c. endemic species unit

d. evolutionarily significant unit

**7.10f   All of the individuals of the same species kept in zoos in North America, and exchanged for breeding purposes between these zoos, can be considered a**

a. micropopulation

b. metapopulation

c. megapopulation

d. monopopulation

**7.11f   The original concept of a National Zoological Park in the United States was in part the result of concerns over the destruction of which species?**

a. Grey wolf (*Canis lupus*)

b. Brown bear (*Ursus arctos*)

c. American bison (*Bison bison*)

d. Caribou (*Rangifer tarandus*)

**7.12f   The captive breeding programme for endangered species operated in North America zoos uses a symbol consisting of two**

a. tigers

b. elephants

c. rhinoceroses

d. giant pandas

**7.13f** **When close relatives mate their offspring may suffer a reduction in genetic fitness known as**

  a.  inbreeding restraint

  b.  inbreeding suppression

  c.  inbreeding regression

  d.  inbreeding depression

**7.14f** **When breeding rare birds it is sometimes possible to induce a female to lay a second clutch of eggs by removing the first clutch and transferring these eggs to an incubator. This ability to replace the eggs that have been removed is known as**

  a.  double laying and increases fecundity

  b.  double clutching and increases fecundity

  c.  double laying and decreases fecundity

  d.  double clutching and decreases fecundity

**7.15f** **In North America, the captive breeding programmes of the Association of Zoos and Aquariums are called**

  a.  Species Conservation Plans

  b.  Conservation Breeding Plans

  c.  Endangered Species Programmes

  d.  Species Survival Plans

**7.16f** **Tubal ligation is**

  a.  a method of contraception used for female mammals

  b.  a method of contraception used for male birds

  c.  an ART used to increase the survival of embryos

  d.  a type of *in-vitro* fertilisation

**7.17f** **All of the giant pandas (*Ailuropoda melanoleuca*) (Fig. 7.1) housed in zoos outside China**

  a.  have been given to these zoos by the Chinese Government as gifts

  b.  have been loaned to these zoos indefinitely by the Chinese Government at no cost

   c. have been temporarily 'rented' from the Chinese Government and any cubs produced are owned by the zoo in which they were born

   d. have been temporarily 'rented' from the Chinese Government and any cubs produced remain the property of the Chinese Government

**Fig. 7.1.**

**7.18f The process of producing individual animals with identical or almost identical DNA is known as**

   a. inbreeding

   b. cloning

   c. cleaving

   d. duplicating

**7.19f Which of the following captive-bred species have been reintroduced to the wild?**

   i. Przewalski's horse (*Equus ferus przewalskii*)

   ii. European bison (*Bison bonasus*)

   iii. Golden lion tamarin (*Leontopithecus rosalia*)

   iv. Mauritius kestrel (*Falco punctatus*)

v. Western swamp turtle (*Pseudemydura umbrina*)

vi. Chinese alligator (*Alligator sinensis*)

a. i, ii and iii

b. i, iii, iv and v

c. ii, iii, iv and vi

d. All of them

**7.20f The first ever studbook for a wild animal was established in 1923 for the**

a. European bison (*Bison bonasus*)

b. Okapi (*Okapia johnstoni*)

c. Père David's deer (*Elaphurus davidianus*)

d. Przewalski's horse (*Equus ferus przewalskii*)

# Intermediate

**7.1i When population size is small, breeding animals in zoos at a younger age than would be usual in the wild may have the effect of rapidly increasing population size**

a. while reducing the rate of loss of genetic diversity

b. but increasing the rate of loss of genetic diversity

c. without any effect on genetic diversity

d. and increasing the mutation rate

**7.2i The quagga is an extinct**

a. camelid

b. ursid

c. mustelid

d. equid

**7.3i Which of the following felids is of little conservation value?**

a. Amur tiger

b. Leopon

c. Asiatic lion

d. Snow leopard

**7.4i**  **The breeding of rare animals in zoos and their subsequent release to the wild is best described as**

a.  common but rarely successful

b.  rare and not always successful

c.  common and always successful

d.  rare and always successful

**7.5i**  **Young animals that have been born and reared in a zoo may be cared for and released to the wild after they have matured to the point where they are less susceptible to predation, starvation or other causes of mortality. This process is known as**

a.  preconditioning

b.  conditioning

c.  headstarting

d.  maturation

**7.6i**  **The modern approach to conservation that integrates species conservation planning by considering all populations (*in-situ* and *ex-situ*) under all management conditions and engages with all responsible parties and resources is called the**

a.  One Plan Approach

b.  One Scheme Approach

c.  Master Plan Approach

d.  One Objective Approach

**7.7i**  **The world's first test tube baby gorilla was born in 1995 to an individual in**

a.  the Bronx Zoo

b.  London Zoo

c.  Cincinnati Zoo

d.  Berlin Zoo

**7.8i**   **Cloning has been criticised as a conservation technique for animals because**

    a.  it is very expensive

    b.  cloned populations lack genetic diversity

    c.  cloned individuals are prone to neonatal health problems

    d.  all of the above are true

**7.9i**   **A keeper might use a candle test to determine whether or not**

    i.   a female seahorse was pregnant

    ii.  a male seahorse was pregnant

    iii. a duck's egg contains a live embryo

    a.  i and iii

    b.  ii only

    c.  ii and iii

    d.  iii only

**7.10i**   **Which of the following organisations keeps the international studbooks for rare and endangered species?**

    a.  IUCN

    b.  WAZA

    c.  WWF

    d.  UNEP

**7.11i**   **When selecting a male tarantula (Theraphosidae) for breeding it is important to ensure that**

    a.  the most caudal pair of legs do not have the tarsus missing

    b.  the first pair of legs are intact

    c.  the pedipalps are intact

    d.  all of the above are true

**7.12i**   **The species in Fig. 7.2 was saved from extinction by captive breeding. It is**

    a.  a Hawaiian goose (*Branta sandvicensis*)

    b.  an Orinoco goose (*Neochen jubata*)

c. a Magellan goose (*Chloephaga picta*)

d. an Andean goose (*Chloephaga melanoptera*)

**Fig. 7.2.**

### 7.13i Which of the following statements is false?

a. Hybrids that have resulted from matings between orangutans from Sumatra and orangutans from Borneo are of low conservation value

b. There are separate zoo-based captive breeding programmes for the different subspecies of Asian elephants

c. There are no mountain gorillas kept in zoos

d. Historically, cheetahs have not bred well in captivity

### 7.14i The ASMP is the

a. Australasian Species Management Program

b. African Species Management Project

c. Asian Species Maintenance Programme

d. American Species Monitoring Project

### 7.15i The computer program developed by scientists at Lincoln Park Zoo, Chicago, to track individual animals through their lifetimes and assist in population management is called

a. *PopStat*

b. *ZooPop*

  c. *PopLink*

  d. *ZooLink*

**7.16i** **The genetic diversity within two inbred zoo populations of the same species (A and B) may be increased if individuals from population A are allowed to interbreed with those from population B due to the phenomenon known as**

  a. hybrid vitality

  b. hybrid vigour

  c. hybrid robustness

  d. hybrid stability

**7.17i** **Sperm sex-sorting techniques can be used to skew the sex ratio within a population to produce more female offspring in species that have female-dominated social groups (e.g. gorillas). Sex-sorting of spermatozoa can be achieved using**

  a. chromatography

  b. differential centrifugation

  c. electrophoresis

  d. flow cytometry

**7.18i** **Vitrification is a type of**

  a. artificial insemination

  b. cryopreservation

  c. *in-vitro* fertilisation

  d. ovulation induction

**7.19i** **The purpose of lifetime reproductive planning for a species is to balance which of the following?**

  i. Early reproduction

  ii. Genetic diversity

  iii. Regular reproduction

  iv. Population age structure

  v. The holding capacity of zoos

a. The benefits of iii with the effects of these on iv and v

b. The benefits of i and iii with the effects of these on ii, iv and v

c. The benefits of i with the effects of these on ii and v

d. The benefits of i and ii with the effects of these on iii and iv

**7.20i MGA (melengestrol acetate) implants are used in zoo animals**

a. as contraception

b. to induce parturition

c. to prevent spontaneous abortion

d. to stimulate ovulation

# Advanced

**7.1a Match the names of the gene banks in Table 7.1 with their locations.**

a. A

b. B

c. C

d. D

Table 7.1

| Gene bank | A | B | C | D |
|---|---|---|---|---|
| Frozen Ark Project | University of Newcastle, New South Wales, Australia | University of Nottingham, United Kingdom | Smithsonian Conservation Biology Institute, United States | University of Newcastle, New South Wales, Australia |
| FrogBank | Leibniz Institute for Zoo and Wildlife Research, Germany | University of Newcastle, New South Wales, Australia | University of Nottingham, United Kingdom | Leibniz Institute for Zoo and Wildlife Research, Germany |
| Genome Resource Bank | University of Nottingham, United Kingdom | Leibniz Institute for Zoo and Wildlife Research, Germany | University of Newcastle, New South Wales, Australia | Smithsonian Conservation Biology Institute, United States |
| The Center for Species Survival's Genome Resource Bank | Smithsonian Conservation Biology Institute, United States | Smithsonian Conservation Biology Institute, United States | Leibniz Institute for Zoo and Wildlife Research, Germany | University of Nottingham, United Kingdom |

**7.2a**  **Sperm used in artificial insemination is most likely to be stored at**

    a. −196°C

    b. 0 °C

    c. −19 °C

    d. −169 °C

**7.3a**  **It is possible to induce a female mammal to mature and release more eggs than usual by treating her with a variety of hormones. This phenomenon is known as**

    a. maxiovulation

    b. hyperovulation

    c. superovulation

    d. multiovulation

**7.4a**  **The AZA Wildlife Contraception Center is located at**

    a. San Antonio Zoo

    b. San Francisco Zoo

    c. San Diego Zoo

    d. St Louis Zoo

**7.5a**  **A GnRH implant in a female chimpanzee would be used to**

    a. control aggression

    b. inhibit oestrus

    c. increase egg production

    d. prevent spontaneous abortions

**7.6a**  **The world's first interspecies frozen/thawed embryo transfer resulted in the cloning of**

    a. an African wild cat (*Felis silvestris lybica*)

    b. a fennec fox (*Vulpes zerda*)

    c. a serval (*Leptailurus serval*)

    d. a cheetah (*Acinonyx jubatus*)

**7.7a** There is sufficient difference between the Himalayan and Chinese populations of the red panda (*Ailurus fulgens*) (Fig. 7.3) for them to be considered to belong to two separate

Fig. 7.3.

    a. ESUs

    b. genera

    c. families

    d. subgenera

**7.8a** **Which of the following organisations have been involved in the Black-Footed Ferret Species Survival Plan® since the recovery programme for this species began?**

    i. Phoenix Zoo, Toronto Zoo and Henry Doorly Zoo

    ii. US Fish and Wildlife Service

    iii. Wyoming State Game and Fish

    iv. Cheyenne Mountain Zoo and Louisville Zoological Gardens

    v. Audubon Nature Institute

    vi. Smithsonian Conservation Biology Institute

    a. i, iv and v

    b. ii, iii, iv and vi

    c. ii, iii and iv

    d. i, ii, iii, iv and vi

**7.9a** Semen collection and artificial insemination have been studied in cockatiels (*Nymphicus hollandicus*) to assist in the development of artificial insemination in psittacines (parrots). In this context the cockatiel was being used as a

a. replica species

b. substitute species

c. prototype species

d. imitation species

**7.10a** When studying the genetic makeup of species and subspecies of conservation interest scientists examine specific sequences of DNA bases to determine relatedness between DNA samples. These sequences are known as

a. minisatellite markers

b. microsatellite markers

c. macrosatellite markers

d. nanosatellite markers

**7.11a** The population of mountain gorillas (*Gorilla beringei beringei*) has made a remarkable recovery as a result of

a. a conventional zoo breeding programme

b. a zoo breeding programme using ARTs

c. the use of veterinary and other extreme interventions *in-situ*

d. a natural recovery of the wild population

**7.12a** Which was the first endangered species to be cloned?

a. Gaur (*Bos gaurus*)

b. Banteng (*Bos javanicus*)

c. Addax (*Addax nasomaculatus*)

d. Scimitar-horned oryx (*Oryx dammah*)

**7.13a** Barbary doves (*Streptopelia risoria*) are used by zoos to rear chicks of the much rarer pink pigeon (*Columba mayeri*) to increase the fecundity of the latter, in a process called

   a.  cross-fostering

   b.  cross-breeding

   c.  cross-fertilisation

   d.  cross-rearing

**7.14a** Complete the following sentence from the list of words below: 'Reproductive rate is increased in some endangered mammal species by implanting embryos in the reproductive tracts of other closely related species. For example, the domestic horse has been used as a ................ species for Przewalski's horse (*Equus ferus przewalskii*) (Fig. 7.4).'

Fig. 7.4.

   a.  proxy

   b.  replacement

   c.  substitute

   d.  surrogate

**7.15a According to the guidelines of the IUCN, which of the following is the correct sequence of actions required when considering and then undertaking a conservation translocation, for example, the reintroduction of zoo-bred animals to the wild (Table. 7.1)?**

Table 7.1

| Sequence | A | B | C | D |
|---|---|---|---|---|
| 1 | Decision to translocate | Goal | Goal | Evaluation of alternatives |
| 2 | Design | Decision to translocate | Evaluation of alternatives | Decision to translocate |
| 3 | Goal | Design | Decision to translocate | Goal |
| 4 | Evaluation of alternatives | Evaluation of alternatives | Design | Design |
| 5 | Implementation | Monitoring | Implementation | Monitoring |
| 6 | Outcome assessment | Implementation | Monitoring | Implementation |
| 7 | Monitoring | Outcome assessment | Outcome assessment | Outcome assessment |

    a. A

    b. B

    c. C

    d. D

**7.16a Which of the following statements about the reintroduction of zoo animals into the wild is false?**

    a. It should be conducted in accordance with the IUCN/SSC/ Reintroduction Specialist Group guidelines for reintroduction

    b. It should be supervised by the Zoological Society of London

    c. It may need a licence from the appropriate government authority

    d. It should be followed by a period of monitoring

**7.17a** **The Center for Reproduction of Endangered Wildlife (CREW) was founded in 1981 at**

a. Cincinnati Zoo

b. Bronx Zoo

c. Philadelphia Zoo

d. San Diego Zoo

**7.18a** **Individual animals in some conservation breeding programmes have a contraceptive implant under the skin. Which of the following would not normally be a legitimate reason for doing this?**

a. Their genes are over-represented in the captive population

b. They have recently been identified as hybrids between two subspecies, each of which has its own breeding programme

c. To delay reproduction when suitable zoo accommodation is unavailable

d. To allow selective breeding for tameness

**7.19a** **The target for maintaining genetic diversity in captive populations of animals is widely accepted to be**

a. 95% of genetic diversity in a demographically stable population for 100 years

b. 85% of genetic diversity in a demographically stable population for 150 years

c. 95% of genetic diversity in a demographically stable population for 200 years

d. 85% of genetic diversity in a demographically stable population for 300 years

**7.20a** **The acronym SCNT refers to**

a. somatic cell nuclear transfer

b. sex cell nuclear transfer

c. single cell nuclear transfer

d. somatic cell nucleus transmission

# 8 Behaviour, Training and Environmental Enrichment

This chapter contains questions on the behaviour of animals in zoos, the techniques used to train them and the environmental enrichment of their enclosures.

## Foundation

**8.1f** A list and description of all – or a subset – of the behaviours a species exhibits is known as

   a.  a sociogram

   b.  an ethnogram

   c.  an ethogram

   d.  an ecogram

**8.2f** Complete the following definition of environmental enrichment suggested by Shepherdson (1998) using one of the options below:

[Environmental enrichment is] *an animal husbandry principle that seeks to enhance the quality of captive animal care by identifying and providing the environmental stimuli necessary for optimal ...............*

   a.  *nutritional and psychological status*

   b.  *psychological and physiological well-being*

   c.  *psychological well-being*

   d.  *physiological functioning*

© Paul A. Rees 2021. *Key Questions in Zoo and Aquarium Studies: A Study and Revision Guide* (P.A. Rees)
DOI: 10.1079/9781789249002.0008

**8.3f** 'Having no flight tendency with respect to humans' is a definition of

a. tameness

b. nervousness

c. aggression

d. none of the above

**8.4f** In the wild, different populations of a forest primate may differ in what they eat due to differences in

a. food tolerances

b. appetency

c. cultural traditions

d. dietary requirements

**8.5f** Which of the following species communicates using sounds that have been given the following names by zoologists: lip smack, soft grunt, bark, scream, nest grunt and tooth clack?

a. Chimpanzees (*Pan troglodytes*)

b. Grey wolves (*Canis lupus*)

c. Jaguars (*Panthera onca*)

d. Common squirrel monkey (*Saimiri sciureus*)

**8.6f** Behavioural engineering in zoos was pioneered by

a. Prof. Heini Hediger

b. Dr Hal Markowitz

c. Carl Hagenbeck

d. Sir Stamford Raffles

**8.7f** Match the terms in Table 8.1 with the time of day when an animal is active

a. A

b. B

c. C

d. D

Table 8.1

| Active | A | B | C | D |
|---|---|---|---|---|
| at dusk or early evening | Cathemeral | Matutinal | Vespertine | Vespertine |
| at twilight | Matutinal | Vespertine | Crepuscular | Crepuscular |
| intermittently throughout the day and night | Vespertine | Crepuscular | Matutinal | Cathemeral |
| at dawn or early morning | Crepuscular | Cathemeral | Cathemeral | Matutinal |

**8.8f    Enrichment allows animals kept in zoos to**

    a.   exhibit natural (species-typical) behaviours

    b.   exercise control or choice over their environment

    c.   experience enhanced wellbeing

    d.   do all of the above

**8.9f    Gustatory enrichment involves the sense of**

    a.   sight

    b.   hearing

    c.   taste

    d.   touch

**8.10f   A fruitsicle is**

    a.   a plastic enrichment toy shaped like a piece of fruit

    b.   a block of ice made of fruit juice and pieces of fruit

    c.   an enrichment device that dispenses pieces of fruit at random times

    d.   a training device that delivers fruit as a reward

**8.11f   A large, almost indestructible plastic ball (Fig. 8.1) given to animals in zoos, especially large carnivores, as an enrichment is a**

    a.   *Brawny Ball*

    b.   *Bully Ball*

    c.   *Boomer Ball*

    d.   *Bailey Ball*

Fig. 8.1.

**8.12f** **The use of complex feeders requiring tongue use to access food is most likely to be beneficial in reducing oral stereotypies in**

a. elephants (Elephantidae)

b. giraffes (*Giraffa camelopardalis*)

c. brown bears (*Ursus arctos*)

d. spider monkeys (Atelidae)

**8.13f** **A behaviour performed by an animal living in a zoo in the absence of the appropriate stimulus, for example a weaver bird (Ploceidae) exhibiting nest-building behaviour when no nesting material is present, is known as**

a. an altruistic behaviour

b. a pointless behaviour

c. a vacuum behaviour

d. an appetitive behaviour

**8.14f** **In the study of the behaviour of animals living in zoos the acronym SIB means**

a. self-injurious behaviour

b. self-inflicted behaviour

c. social interaction behaviour

d. semi-intentional behaviour

**8.15f** **Elephants are sometimes handled and trained using a tool consisting of a metal hook attached to a handle (Fig. 8.2). This tool is called**

a. a goad

b. an ankus

c. a bullhook

d. all of the above

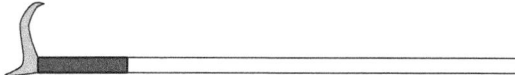

Fig. 8.2.

**8.16f** **The training of an animal to perform an act by progressively manipulating its behaviour by rewarding activity in the direction of the desired act is called**

a. moulding

b. shaping

c. framing

d. guiding

**8.17f** **A form of territory held by the males of certain species and used solely as a communal mating ground – especially in birds such as grouse, pheasants and bowerbirds – where females select mates is known as**

a. a lek

b. a patch

c. a plot

d. a tract

**8.18f** Animals that have been introduced to a new enclosure may initially display an escape response each time a visitor appears in view. Eventually they will learn that the visitors pose no threat and this response will diminish and may eventually disappear completely. This type of learning is called

a. adaptation

b. conditioning

c. imprinting

d. habituation

**8.19f** Which of the following statements about ethograms is false?

a. An ethogram must include all the known behaviours of a species

b. The definitions of the behaviours in an ethogram should be mutually exclusive

c. When calculating an activity budget it is legitimate to pool behaviours from several categories if they have low frequencies

d. The best ethograms are functional (i.e. the categories are meaningful to the animal)

**8.20f** Which of the following statements about pheromones is false?

a. They may be used as sexual attractants

b. They are used by territorial animals to mark their territories

c. They are not known in aquatic animals

d. They are detected by the recipient's olfactory system

# Intermediate

**8.1i** Some birds, such as geese, produce chicks that have their eyes open as soon as they hatch and are instantly capable of running or swimming away from threats such as predators. Such chicks are referred to as

a. altricial

b. intricate

c. refined

d. precocial

**8.2i** **The inability of an animal living in a zoo to perform its full repertoire of behaviours is called**

a. behavioural limitation

b. behavioural restriction

c. behavioural confinement

d. behavioural adaptation

**8.3i** **The red spot test is used to test for**

a. tuberculosis

b. self-recognition

c. foot-and-mouth disease

d. skin allergies

**8.4i** **The modification of voluntary behaviour by the use of a reward is called**

a. operant conditioning

b. imprinting

c. reinforcement

d. classical conditioning

**8.5i** **Which of the following animals are not capable of saltatory locomotion?**

a. Kangaroos

b. Sloths

c. Frogs

d. Lemurs

**8.6i** **Which of the following is not an important element of handling elephants using protected contact?**

a. One or more targets

b. A food reward

    c. An ankus

    d. A metal fence between the keepers and the elephants

**8.7i** **Complete the following sentence using one of the terms listed below: 'Animals may be considered to possess ............... when individuals differ from one another in either single behaviours or suites of related behaviours in a way that is consistent over time'.**

    a. a personality

    b. sentience

    c. a character

    d. consciousness

**8.8i** **A movement pattern that is performed repeatedly, relatively invariant in form, and has no apparent function or goal is called**

    a. a ritualised behaviour

    b. an imprinted behaviour

    c. a deimatic display

    d. a stereotypy

**8.9i** **Hyperaggression is exhibited by some species in captivity. This change in behaviour compared with what would be considered a normal level of aggression is**

    a. a temporal change

    b. a qualitative change

    c. a quantitative change

    d. an empirical change

**8.10i** **Instrumental learning may be used in zoos to assist with**

    a. moving animals between enclosures

    b. veterinary examinations

    c. transportation of animals between zoos

    d. all of the above

**8.11i** The straps labelled 'X' in Fig. 8.3 are used in the tethering and training of birds of prey and are called

a. tethers

b. ties

c. jesses

d. thongs

Fig. 8.3.

**8.12i** When animals such as chimpanzees (*Pan troglodytes*) voluntarily use touch screen computers to communicate with researchers or participate in research by undertaking mental tasks they are engaging in

a. analytical enrichment

b. logical enrichment

c. empirical enrichment

d. cognitive enrichment

**8.13i** Which of the following species is most likely to use an enrichment device known as a 'wobble tree'?

a. A green python (*Morelia viridis*)

b. A sloth bear (*Melursus ursinus*)

c. A nine-banded armadillo (*Dasypus novemcinctus*)

d. A red-legged seriema (*Cariama cristata*)

**8.14i  A piñata is**

    a.  a container made from papier-mâché, cardboard, cloth or similar material containing food treats

    b.  a type of fruit eaten by fruit bats indigenous to Indonesia

    c.  an element of the courtship behaviour of some tropical frogs

    d.  a structure built by a male a bird of paradise to attract a mate

**8.15i  The relationships between individuals in a social group of Asian elephants (*Elephas maximus*) are shown in Fig. 8.4. This type of diagram is called**

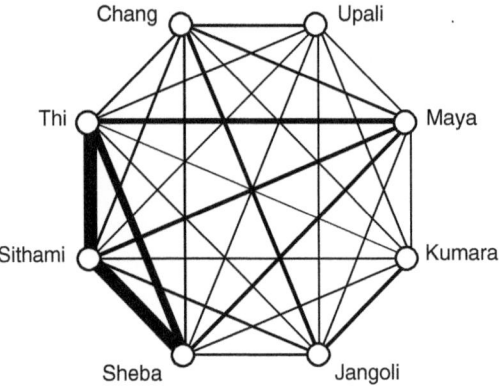

**Fig. 8.4.**

    a.  an ethogram

    b.  a cluster analysis

    c.  a sociogram

    d.  a Venn diagram

**8.16i  In Fig. 8.4 the thickness of the line joining any two individuals represents the degree of association recorded between them. Which elephant is the least social?**

    a.  Chang

    b.  Upali

    c.  Maya

    d.  Sheba

**8.17i** During the training of an animal, a signal that marks the point in time when the animal successfully performs the desired behaviour – such as the sound of a clicker – is known as the

a. cue

b. prompt

c. link

d. bridge

**8.18i** As part of the process of training Aldabra tortoises (*Geochelone gigantea*) to accept venipuncture the animals were taught to approach a target consisting of a red ball attached to the end of a pole (Weiss and Wilson, 2003). Initially the target was placed 25-50mm from the tortoise's face. If the animal moved towards the target a clicker was sounded and a food reward given. This process is known as

a. classical conditioning

b. imprinting

c. operant conditioning

d. observational learning

**8.19i** In question 8.18i the food reward given to the tortoise functions as

a. positive reinforcement

b. confirmatory reinforcement

c. constructive reinforcement

d. affirmative reinforcement

**8.20i** In searching for the immediate reason why an individual animal behaves in a particular way in response to a specific stimulus we are seeking the

a. ultimate cause

b. proximate cause

c. functional cause

d. imminent cause

# Advanced

8.1a    The aspect of animal care that includes the practical applications of animal training, environmental enrichment and behavioural research may be referred to as

    a. behavioural ecology management

    b. holistic animal husbandry

    c. cognitive animal management

    d. behavioural husbandry management

8.2a    In a study of the activity budget of an antelope, behaviour was recorded at 5 minute intervals over three hours. The results are shown in Table 8.2.

Table 8.2

| Behaviour | No. of samples when behaviour was recorded |
|---|---|
| Feeding | 18 |
| Drinking | 2 |
| Lying down | 8 |
| Walking | 5 |
| Out of sight | 3 |

How would you calculate the percentage time spent feeding if the times when the animal was out of sight were excluded from the total study time?

    a. 100 x (18-3/36)

    b. (18/36) x 100

    c. (18/(36-3)) x 100

    d. ((18/36)-3) x 100

8.3a    Which of the following enrichment devices would be especially suitable for use by a bison (*Bison* sp.) to encourage natural behaviour?

    a. Popsicle

    b. Head-butting post

c. Burlap bag

d. Puzzle feeder

**8.4a** **Animals may use different parts of their enclosures unequally as a result of**

i. inadequacies in enclosure design

ii. differences in the basic biology of the species

iii. preferences for, or aversions to, different structural features

iv. attraction to, or fear of, visitors

a. i, ii and iii

b. i, ii and iv

c. i, iii and iv

d. i, ii, iii and iv

**8.5a** **An elephant may be taught to lift its feet through a gate in a steel fence by target training. Which of the following statements is false?**

a. If the elephant stands in the wrong place because the keeper makes a mistake the elephant should still be rewarded

b. Two targets are usually needed for this type of training

c. Target training is a type of classical conditioning

d. Very small pieces of food are suitable as a reward

**8.6a** **Perseveration is the carrying on of an activity by an animal**

a. repeatedly until the desired result is achieved

b. in the absence (or after the cessation) of the appropriate stimulus

c. to regulate its body temperature

d. that it has been trained to perform by operant conditioning

**8.7a** **The term 'zoochosis' was first coined in 1992 by**

a. Bill Travers

b. Terry Maple

c. Georgia Mason

d. Donald Broom

**8.8a**   **A fear of new objects is called**

  a.  xenophobia

  b.  necrophilia

  c.  neophilia

  d.  neophobia

**8.9a**   **The elephant in Fig. 8.5 is reaching out of his enclosure to forage on plants even though food is available in the enclosure itself. This type of behaviour – unnecessarily working for food – is seen in a number of captive animals and is known as**

  a.  freeloading

  b.  contrafreeloading

  c.  opportunism

  d.  pseudofreeloading

Fig. 8.5.

**8.10a  SPIDER is an acronym used in**

    a.  studies of environmental enrichment

    b.  animal training

    c.  behaviour recording

    d.  studies of arachnids

**8.11a  Table 8.3 is a matrix of dominance relationships in a group of four animals A-D. For each animal the matrix shows the number of times it supplanted another animal and the number of times it was supplanted by another individual at a food source. For example, A did not supplant B at all, but supplanted C 12 times and D 28 times. A was supplanted by B 24 times but not supplanted by either C or D. Which was the most dominant animal in this group?**

Table 8.3

| | | Number of times individual was supplanted | | | |
| --- | --- | --- | --- | --- | --- |
| | | A | B | C | D |
| Number of times individual supplanted another | A | | 0 | 12 | 28 |
| | B | 24 | | 17 | 9 |
| | C | 0 | 0 | | 0 |
| | D | 0 | 0 | 7 | |

    a.  A

    b.  B

    c.  C

    d.  D

**8.12a  The course of development of a particular behaviour is known as its**

    a.  phylogeny

    b.  physiognomy

    c.  phrenology

    d.  ontogeny

**8.13a** The ability of an animal to exhibit appropriate behaviour in a given set of circumstances is known as

    a. behavioural capacity

    b. behavioural proficiency

    c. behavioural competence

    d. psychological ability

**8.14a** The males of some antelope species exhibit a leg beat as an element of their

    a. courtship behaviour

    b. appeasement behaviour

    c. aggressive behaviour

    d. parenting behaviour

**8.15a** The opposite of habituation is

    a. adaptation

    b. sensitisation

    c. desensitisation

    d. adjustment

**8.16a** A gorilla (*Virunga*) learned that her keeper (Aisha), always hid small pieces of apple in the same section of her enclosure, in the grass near the climbing frame. Each morning when *Virunga* was allowed access to her outdoor enclosure she ran straight to the climbing frame to gather up and eat the apple. When Aisha moved to another zoo the new gorilla keeper did not continue the practice of hiding food in the enclosure. Initially, each morning *Virunga* continued to search for apple near the climbing frame but without success. Eventually she stopped looking. The behaviour had become

    a. extinct

    b. deleted

    c. rescinded

    d. redacted

**8.17a** **A lizard chases, catches and then eats a live insect in his vivarium. This is an example of**

a. consummatory behaviour followed by appetitive behaviour

b. appetitive behaviour followed by goal-directed behaviour

c. stereotypic behaviour followed by consummatory behaviour

d. appetitive behaviour followed by consummatory behaviour

**8.18a** **Animals in captivity often engage in more play behaviour than their wild conspecifics because**

a. play is a leisure activity and captivity frees up time that wild animals spend on behaviours linked to survival, for example foraging and vigilance behaviour

b. engaging in play in captivity is the result of boredom and lack of stimulation

c. many species in captivity do not develop normal adult behaviour

d. enrichment devices provide more opportunity for play

**8.19a** **In reptiles brumation is**

a. a seasonal change in feeding behaviour

b. a seasonal change in appearance

c. a period of dormancy that occurs in very cold weather

d. a period of dormancy that occurs in very hot weather

**8.20a** **Advertisement, mate attraction, mate selection, mate assessment and sexual coordination are all functions of**

a. courtship

b. oestrus synchrony

c. progesterone

d. mate guarding

# 9 Animal Welfare and Conservation Medicine

This chapter contains questions on the welfare of animals in zoos, the veterinary care of these animals and some common diseases and conditions.

## Foundation

**9.1f** The foundation of any medical programme for zoo animals should be

   a.  preventative medicine

   b.  regular medication

   c.  surgical intervention

   d.  rapid diagnosis

**9.2f** With respect to animals living in zoos the comparable life argument states that

   a.  we should not compare animals with people when thinking about whether or not they are happy living in a zoo

   b.  a zoo should only keep animals in the same enclosure if they have comparable ecological requirements

   c.  it is wrong to keep a species in a zoo if its life in the zoo is not at least as good as it would be in the wild

   d.  conditions in zoos and in the wild are comparable for some species

© Paul A. Rees 2021. *Key Questions in Zoo and Aquarium Studies: A Study and Revision Guide* (P.A. Rees)
DOI: 10.1079/9781789249002.0009

**9.3f** Which of the following statements is false? All animals added to a zoo collection should

a. be introduced to the social group immediately

b. normally be held in quarantine for a suitable period

c. tested for parasites

d. screened for disease

**9.4f** When a sick animal will not eat, has diarrhoea and a fever, these are

a. signs

b. symptoms

c. a combination of signs and symptoms

d. neither signs nor symptoms

**9.5f** A veterinary surgeon decides to euthanise an animal that appears to be perfectly healthy. This is

a. only justifiable for population control

b. sometimes justifiable for a variety of reasons

c. only justifiable if the animal is very old

d. never justifiable

**9.6f** A shift box is most likely to be used in the handling of

a. a venomous snake

b. a cichlid

c. a giraffe

d. an ostrich

**9.7f** Malocclusions are common dental problems found in mammals kept in zoos, that is, they have

a. missing teeth

b. diseased teeth and gums

c. misaligned jaws

d. misaligned teeth or jaws

**9.8f** **To facilitate animal identification and the accurate keeping of veterinary records some animals are fitted with a small electronic device called**

a. an intramuscular transducer

b. a subcutaneous capacitor

c. a subcutaneous transponder

d. an interventricular transponder

**9.9f** **Complete the following sentence with one of the terms listed below: 'When an animal – especially an ungulate – experiences stress as a result of being chased and captured it may experience capture myopathy whereby ......... accumulates in its muscles, sometimes resulting in sudden death'.**

a. carbon dioxide

b. lactic acid

c. calcium

d. potassium

**9.10f** **An analgesic drug is used to**

a. prevent ovulation

b. treat bacterial infections

c. treat cancer

d. relieve pain

**9.11f** **Morbidity is a measure of the incidence of**

a. adult deaths

b. neonate deaths

c. disease

d. infertility

**9.12f** *Jake* **the orangutan has had diabetes for 11 years. His condition is best described as**

a. acute

b. chronic

    c. asymptomatic

    d. persistent

**9.13f The spectacled bear (*Tremarctos ornatus*) in Fig. 9.1 is exhibiting**

    a. route-tracing behaviour

    b. stereotypic behaviour

    c. abnormal behaviour

    d. all of the above

Fig. 9.1.

**9.14f Which of the following could be used to examine the weight distribution on the feet of a large mammal during locomotion?**

    a. An ultrasound scanner

    b. Pressure platforms

    c. An accelerometer

    d. A weigh bridge

**9.15f A young giraffe called *Marius* was euthanised at Copenhagen Zoo in 2014. This event attracted international press attention because**

    a. he had a broken leg and there was no possibility of recovery

    b. he was injured as a result of mishandling

c. he was healthy but his genes were over-represented in the captive population

d. he had seriously injured a keeper and was considered too dangerous for staff to handle

**9.16f  Which of the following statements about stereotypic behaviour is false?**

a. There is no evidence of physical changes to the brain in animals that exhibit stereotypic behaviour

b. Stereotypic behaviour is not exclusively seen in zoos

c. It is sometimes reversible

d. Many scientists believe that it is an indicator of poor welfare

**9.17f  Which of the following statements about longevity are true?**

i. Individuals of some species kept in zoos live longer lives than their wild conspecifics

ii. Individuals of some species kept in zoos live shorter lives than their wild conspecifics

iii. It will be many years before it will be possible to say if improvements to animal welfare in zoos have caused any increase in longevity in captive born individuals of long-lived species

iv. Longevity is the length of an animal's life from birth (or hatching) to death

v. Longevity, lifespan and life expectancy all mean the same thing.

a. ii, iv and v

b. i, ii and iv

c. i, iii, iv and v

d. i, ii, iii and iv

**9.18f  Most reptiles need to be exposed to UVB light to prevent**

a. gastro-intestinal disease

b. metabolic bone disease

c. liver disease

d. reproductive failure

9.19f Zoo veterinary staff have a number of options for immobilis-
ing a large mammal with a drug injected using a hypodermic
syringe depending upon how close they are able to approach
it. Which column in Table 9.1 indicates the most appropriate
method as distance to the target animal increases?

 a. A

 b. B

 c. C

 d. D

Table 9.1

| | A | B | C | D |
|---|---|---|---|---|
| Increasing distance to target | Dart blowpipe | Pole syringe | Pole syringe | Pole syringe |
| | Pole syringe | Dart blowpipe | Dart blowpipe | Dart pistol |
| | Dart pistol | Dart rifle | Dart pistol | Dart blowpipe |
| | Dart rifle | Dart pistol | Dart rifle | Dart rifle |

9.20f Over-supplementation of the food given to lizards with calcium
can result in

 a. hypocalcaemia

 b. hypercalcaemia

 c. hypoglycaemia

 d. hyperhidrosis

# Intermediate

9.1i Which of the following taxa may suffer from bumblefoot
(pododermatitis)?

 i. Penguins

 ii. Flamingos

 iii. Rodents

 iv. Rabbits

a. i and ii

b. iii and iv

c. i, ii and iii

d. i, ii, iii and iv

**9.2i    Dystocia is**

a. a disease of the gut

b. an abnormality of the jaw

c. a difficult birth

d. an abnormality of the foetus

**9.3i    Routine cleaning of the teeth and the provision of chewing materials can help to prevent and treat**

a. malocclusions

b. diastema

c. oedema

d. periodontal disease

**9.4i    Psittacosis is a disease found in birds and mammals, including humans, caused by a**

a. virus

b. bacterium

c. prion

d. cestode

**9.5i    In a paper published in 2005 Goossens *et al.* reported the results of 'A 12-month survey of the gastrointestinal helminths of antelopes, gazelles and giraffids kept at two zoos in Belgium.' What type of parasites were examined in this work?**

a. Bacteria

b. Protozoans

c. Flatworms

d. Nematodes

**9.6i    West Nile virus can infect**

a.  birds

b.  mammals

c.  reptiles

d.  all of the above

**9.7i    Spondylosis is a common condition in old bears (Ursidae) and affects the**

a.  heart

b.  vertebrae

c.  leg joints

d.  muscles

**9.8i    Which of the following could not suffer from or carry rabies?**

a.  A fruit bat

b.  A skunk

c.  A monitor lizard

d.  A squirrel monkey

**9.9i    A nebuliser is an**

a.  aerosol delivery system used to deliver some drugs

b.  a type of anaesthetic

c.  a device used to stabilise a broken limb

d.  a type of antiseptic

**9.10i    Chronic stress in zoo animals may lead to**

a.  death

b.  lethargy

c.  depression

d.  any of the above

**9.11i** **Captive octopuses sometimes eat their own arms in a self-mutilation process known as**

a. autotilly

b. autophagy

c. autotomy

d. autolysis

**9.12i** **An animal that experiences emesis as a result of transportation has**

a. vomited

b. collapsed

c. become disorientated

d. entered a state of torpor

**9.13i** **Laminitis is a condition that affects**

a. the heart in primates

b. the feet in ungulates

c. the skin in reptiles

d. the gills in fish

**9.14i** **Which of the following is not a drug used to immobilise large mammals such as a big cat?**

a. Ketamine

b. Etorphine

c. Revivon

d. Xylazine

**9.15i** **Which of the following statements about a body condition scoring system is false?**

a. It should take into account age

b. It should be species specific

c. A body condition score is always between 1 and 5

d. It should be independent of weight

**9.16i** **Which of the following statements about foot-and-mouth disease is false?**

a. Infected animals have small fluid-filled blisters in the mouth and on the feet

b. It is highly contagious

c. It can affect all cloven-hoofed species

d. The disease cannot be transmitted to humans

**9.17i** **The concentration of which of the following may be used as an indicator of stress?**

a. Cholesterol

b. Cortisol

c. Cholecystokinin

d. Cytosine

**9.18i** **A narrow metal enclosure used for restraining livestock while individuals are vaccinated, treated, branded or for some other reason is called a**

a. crush

b. chute

c. shift

d. rack

**9.19i** **Feline panleukopenia is a highly contagious disease that**

a. is caused by a parvovirus and affects felids and related families of mammals

b. is caused by a parvovirus and only affects felids and canids

c. is caused by a prion and only affects felids

d. is caused by a bacterium and only affects felids and mustelids

**9.20i** **Excessive tail bobbing (up and down movements of the tail) is a sign of respiratory distress in**

a. felids

b. elephants

c. small birds

d. lizards

# Advanced

9.1a   A leading cause of disease and death in macropods living in zoos presents as inflammation of the jaw bone with characteristic abscesses and is called

a. bumpy jaw

b. lumpy jaw

c. chunky jaw

d. hunky jaw

9.2a   An infectious animal disease that poses a serious threat to agricultural and other animals, and in some cases humans, whose occurrence must by law be reported to the government, is known in many counties as a

a. disclosure disease

b. communicable disease

c. transmissible disease

d. notifiable disease

9.3a   In most species coprophagia is an unnatural behaviour in which an animal

a. refuses to eat

b. defecates repeatedly

c. eats faeces

d. eats excessively

9.4a   During the transfer of some large mammals between zoos the transportation equipment includes a means of controlling the climate inside the animal crate called a

a. HMAC system

b. HVAC system

    c. TVAC system

    d. HVAD system

**9.5a** ***Live Hard, Die Young*** **is a report by the Royal Society for the Prevention of Cruelty to Animals (RSPCA) on the welfare of which animals in zoos?**

    a. Polar bears

    b. Killer whales

    c. Elephants

    d. Chimpanzees

**9.6a** **A veterinary report on the condition of a fruit bat said that 'Radiographs revealed luxation of the coxofemoral joint…'. This means that the joint was**

    a. dislocated

    b. suffering from degenerative bone disease

    c. infected

    d. inflamed

**9.7a** **The ostrich in Fig. 9.2 needs to be immobilised for veterinary treatment. Which are the two best places on the bird's body for the vet to aim the hypodermic dart?**

    a. 1 and 2

    b. 3 and 4

    c. 1 and 4

    d. 2 and 3

**Fig. 9.2.**

**9.8a** **Which of the following was the first zoo-housed animal to be diagnosed with COVID-19 in the United States?**

a. A lion at Detroit Zoo

b. A leopard at San Francisco Zoo

c. A tiger at the Bronx Zoo

d. A cheetah at St Louis Zoo

**9.9a** **Which of the following frequently develop a collapsed dorsal fin when kept in captivity?**

a. Male orcas (*Orcinus orca*)

b. Female lemon sharks (*Negaprion brevirostris*)

c. Male bottlenose dolphins (*Tursiops* spp.)

d. Female harbour porpoises (*Phocoena phocoena*)

**9.10a** *Ichthyophthirius multifiliis* causes white spot in fish and is

    a. a fungus

    b. a nematode

    c. a ciliated protozoan

    d. an amoeba

**9.11a** **The size of most animal enclosures is not based on any objective measure of what is appropriate for the animals' welfare needs. According to Browning and Maple (2019) which of the following is likely to be the most useful metric of usable exhibit space?**

    a. Usable length

    b. Usable height

    c. Usable area

    d. Usable volume

**9.12a** **Complete the following statement using the words in one of the options listed below: 'Chytridiomycosis is a disease responsible for the global decline in ........ which is caused by a ........'.**

    a. reptiles; fungus

    b. reptiles; virus

    c. amphibians; fungus

    d. amphibians; virus

**9.13a** **An elevator is used to**

    a. remove teeth

    b. clean infected tissue

    c. remove diseased organs

    d. inspect the body cavity

**9.14a The McMaster flotation technique is used to**

    a. determine the digestibility of foodstuffs

    b. examine blood samples for the presence of blood parasites

    c. count parasite eggs in faeces

    d. determine the specific gravity of salt water in a marine aquarium

**9.15a Which of the following statements about stress and distress is false?**

    a. An animal may experience stress without being in distress

    b. Some animals may show no outward behavioural signs of experiencing stress

    c. There is no evidence of intraspecies differences in responses to stressors

    d. Stress reactions are normal reactions to the environment and may be considered adaptive

**9.16a The data illustrated in Fig. 9.3 show the relationship between the mean frequency of stereotypic behaviour in a group of individual large mammals living in a zoo and two variables. These data were collected between 1000h and 1400h on 35 days. Which of the following statements is not supported by the data?**

    a. The mean frequency of stereotypic behaviour increases throughout the day

    b. The animals experienced more stress on cold days than on warm days

    c. For any particular time of day the mean frequency of stereotypic behaviour was almost always higher on cold days than on warm days

    D. On warm days very little stereotypic behaviour was observed between 1000h and 1100h

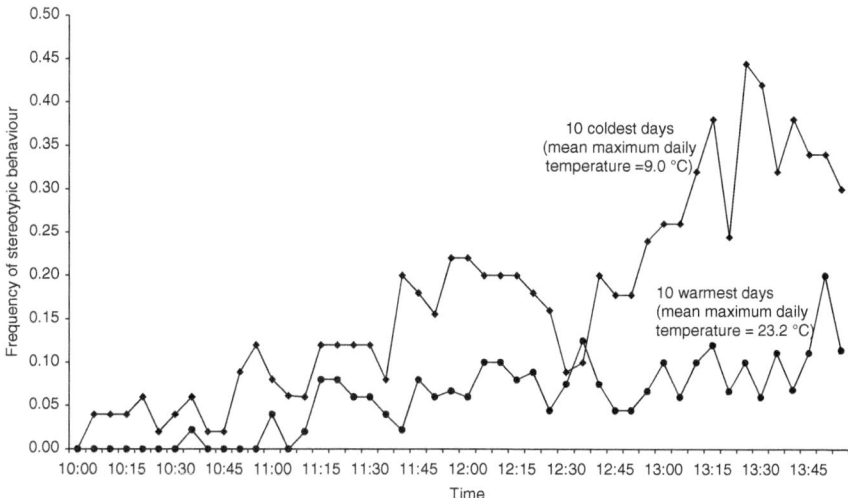

Fig. 9.3.

**9.17a** **Levels of cortisol in the body may be measured from samples of which of the following?**

    i.   Blood

    ii.  Saliva

    iii. Urine

    iv. Faeces

    a.  i, ii and iii

    b.  ii, iii and iv

    c.  iii and iv

    d.  i, ii, iii and iv

**9.18a** **Quirke *et al.* (2012) examined the factors that influenced the prevalence of stereotypical behaviours in captive cheetahs (*Acinonyx jubatus*). Which of the following would you expect to have increased the prevalence of this behaviour?**

    i.   Being solitary

    ii.  Being fed on a predictable feeding regime

    iii. Increased enclosure size

    iv. The ability to see cheetahs in adjacent enclosures

a. i, ii and iii

b. i, iii and iv

c. i, ii and iv

d. ii, iii and iv

**9.19a  Multiple ocular colobomas occur in snow leopards (*Uncia uncia*) (Fig. 9.4) and are**

a. a result of a bacterial infection of the eyes

b. congenital eye malformations

c. caused by cataracts

d. a result of injury to the eye

**Fig. 9.4.**

**9.20a  Which of the following statements about quarantine in zoos and aquariums is false?**

a. The main purpose of quarantine is the prevent the introduction of transmissible disease into a collection

b. There is evidence that very little transmissible disease is transferred by animals moved between accredited zoos

c. All animals transferred to a zoo or aquarium must always be quarantined immediately on arrival

d. Animals obtained from private animal breeders and confiscated animals are more likely to be sources of disease than those obtained from accredited zoos.

# 10 Zoo Organisation and Regulation

This chapter contains questions on the organisation of zoos, the role of zoo professionals, record keeping, zoo legislation and the international conventions that affect zoos.

## Foundation

**10.1f The Zoological Society of London (Fig. 10.1) is based in**

    a. Regent's Park

    b. Richmond Park

**Fig. 10.1.**

© Paul A. Rees 2021. *Key Questions in Zoo and Aquarium Studies: A Study and Revision Guide* (P.A. Rees)
DOI: 10.1079/9781789249002.0010

c. St James's Park

d. Hyde Park

**10.2f The acronym CITES refers to an international treaty that**

a. regulates the organisation of zoos

b. controls trade in endangered species

c. co-ordinates the conservation work of zoos

d. protects the environment

**10.3f The conservation NGO that consists of a network of aquariums, zoos, universities, researchers and government organisations whose aim is to improve animal welfare and species conservation is known as**

a. Species90

b. Species180

c. Species300

d. Species360

**10.4f Which of the following is not an association of zoos?**

a. WAZA

b. BIAZA

c. ABWAK

d. SEAZA

**10.5f Which of the following is least likely to be determined by an endocrinologist working in a zoo?**

a. The amount of stress experienced by snow leopards during transportation between zoos

b. The nutritional value of food eaten by Komodo dragons

c. The time of ovulation in a female gorilla

d. Whether or not a giant panda is pregnant

**10.6f Zoos are usually licensed by**

a. local government

b. state government

c. national government

d. any of the above, depending on the country where the zoo is located

**10.7f  The Leibniz Institute for Zoo and Wildlife Research is located in**

a. Vienna

b. Berlin

c. Hamburg

d. Frankfurt

**10.8f  The acronym TAG stands for**

a. Taxon Advisory Group

b. Threatened Animal Group

c. Threatened Ape Group

d. Turtle Advisory Group

**10.9f  The conservation status of animals and plants is assessed by the IUCN and these assessments are published in its**

a. Blue List

b. Green List

c. Red List

d. Black List

**10.10f  Which of the following is an *in-situ* conservation project?**

a. An Asian lion breeding programme located in a European zoo

b. A chimpanzee breeding project located in a zoo in Kenya

c. A black rhinoceros conservation project in a National Park in Kenya

d. A cichlid fish breeding project in a Canadian aquarium

**10.11f  Which of the following sequences shows the development of international organisations concerned with the recording and dissemination of data on animals kept in zoos in the correct chronological order?**

a. International Species Inventory System → International Species Information System → Species360

b. International Species Information System → International Species Inventory System → Species360

c. International Species Inventory System → Species360 → International Species Information System

d. Species360 → International Species Inventory System → International Species Information System

**10.12f The Association of Zoos and Aquariums (AZA) accredited its first zoo in**

a. 1962

b. 1974

c. 1981

d. 1986

**10.13f A person whose job consists of developing computer models for conservation planning, along with the risk assessment and population management of threatened species is likely to be qualified and experienced in**

a. environmental management

b. conservation biology

c. computational biology

d. information technology

**10.14f Traditionally someone who is responsible for the management of a particular taxon of animals in a zoo (for example mammals or birds) is called the**

a. custodian

b. curator

c. studbook keeper

d. superintendent

**10.15f International trade in the species in Fig. 10.2 is regulated by CITES. For animals, CITES defines a specimen as**

a. any living animal

b. any animal whether alive or dead

c. any recognisable part or derivative of any animal

d. any animal whether alive or dead or any recognisable part or derivative thereof

Fig. 10.2.

**10.16f The AZA accredits zoos and aquariums in**

    a. the United States only

    b. North America only

    c. the Americas

    d. the Americas and several other countries

**10.17f In relation to zoos, the organisation known as IATA sets international standards for**

    a. animal husbandry

    b. conservation breeding

    c. movements of live animals by air

    d. storage of eggs and sperm

**10.18f Which of the following statements about record keeping in zoos is false?**

    a. In the European Union there is no legal requirement for zoos to keep records of their animals

    b. The *International Zoo Yearbook* is a good source of historical records of the numbers of animals kept by zoos

   c. Zoo records often give only a nominal monetary value to each animal

   d. A large zoo may have more than one studbook keeper

**10.19f The acronym EUAC stands for the**

   a. European Union of Aquarium Curators

   b. European United Aquariums Committee

   c. European Underwater and Aquarium Committee

   d. Eurasian Union of Aquarium Curators

**10.20f In total, how many zoos and aquariums were accredited by the AZA in 2020?**

   a. 53

   b. 106

   c. 240

   d. 376

## Intermediate

**10.1i The acronym ZIMS stands for**

   a. Zoo Information Management System

   b. Zoological Index of Mammal Species

   c. Zoo and International Conservation Status

   d. Zoological Information Management System

**10.2i In the European Union a zoo requires a licence if it is open to the public on**

   a. 7 or more days per year

   b. 10 or more days per year

   c. 21 or more days per year

   d. 365 days of the year

**10.3i** A document drawn up by a zoo that shows the species it keeps at the time of writing and the species it intends to keep in the future is called a

  a. zoo plan

  b. collection plan

  c. collection strategy

  d. taxonomic plan

**10.4i** Which of the following zoo staff is most likely to be involved in keeping records of the numbers of each species kept by the zoo?

  a. Curator

  b. Conservation manager

  c. Director

  d. Registrar

**10.5i** *ZooLex* is an organisation concerned with

  a. zoo design

  b. zoo animal nutrition

  c. zoo animal behaviour

  d. zoo visitor studies

**10.6i** Which of the following organisations established a Conservation Grants Fund in 1984 to support its conservation, scientific and educational initiatives?

  a. WAZA

  b. AZA

  c. EAZA

  d. BIAZA

**10.7i** The organisation that carries out inspections of zoos in the United States is the

  a. Unites States Fish and Wildlife Service

  b. Humane Society of the United States

c. Animal and Plant Health Inspection Service

d. Environmental Protection Agency

**10.8i In 2021 the World Association of Zoos and Aquariums (WAZA) estimated that its member organisations are visited annually by more than**

a. 300 million visitors

b. 500 million visitors

c. 700 million visitors

d. 900 million visitors

**10.9i The Wildlife Conservation Society is the parent body of the**

a. San Diego Zoo, California

b. Smithsonian National Zoo, Washington, D.C.

c. Brookfield Zoo, Chicago

d. Bronx Zoo, New York

**10.10i In the European Union, the Balai Directive is concerned with**

a. the nutrition of zoo animals

b. the trade in wild animals

c. the export, import, and movement of live animals and germplasm

d. the storage of the eggs and sperm of endangered species

**10.11i *Committing to Conservation* was published in 2015 and is the conservation strategy of**

a. the European Association of Zoo and Aquaria (EAZA)

b. the World Association of Zoos and Aquariums (WAZA)

c. the British and Irish Association of Zoos and Aquariums (BIAZA)

d. the Association of Zoos and Aquariums (AZA)

**10.12i The Head of Herpetology in a zoo is responsible for the**

 a. amphibians and reptiles

 b. amphibians

 c. reptiles

 d. invertebrates

**10.13i Who appoints zoo inspectors in England?**

 i. The relevant Secretary of State

 ii. The Local Authority in whose jurisdiction the zoo is located

 iii. BIAZA

 iv. Each zoo nominates its own inspectors

 a. i and iv

 b. ii

 c. i and iii

 d. i and ii

**10.14i Approximately how many licensed zoos and aquariums were there in the United Kingdom in 2020?**

 a. 100

 b. 300

 c. 500

 d. 700

**10.15i Which of the following zoo organisations was responsible for the conservation projects listed in Table 10.1?**

 a. EAZA

 b. WAZA

 c. BIAZA

 d. AZA

**Table 10.1**

| Years | Campaign |
|-------|----------|
| 2015-2017 | Let It Grow |
| 2013-2015 | Pole to Pole |
| 2011-2013 | Southeast Asia |
| 2010-2011 | Ape Campaign |
| 2008-2010 | European Carnivore |
| 2007-2008 | Amphibian |
| 2006-2007 | Madagascar |
| 2005-2006 | Rhino |
| 2004-2005 | ShellShock |
| 2002-2004 | Tiger |
| 2001-2002 | Rainforest |
| 2000-2001 | Bushmeat |

**10.16i** **The government guidelines for the proper operation of zoos in England are published in the**

   a.  *Zoos Forum Handbook*

   b.  *Zoos Expert Committee Handbook*

   c.  *Secretary of State's Standards of Modern Zoo Practice*

   d.  *BIAZA Zoos Handbook*

**10.17i** **Which of the following organisations is not anti-zoo?**

   a.  PETA

   b.  Born Free Foundation

   c.  CAPS

   d.  CZA

**10.18i** **Which of the lists in Table 10.2 shows the IUCN Red List categories in the correct sequence in relation to data availability and increasing risk of extinction?**

   a.  A

   b.  B

c. C

d. D

**Table 10.2**

| Category | A | B | C | D |
|---|---|---|---|---|
| Increasing risk of extinction | Not Evaluated | Not Evaluated | Not Evaluated | Not Evaluated |
| | Data Deficient | Data Deficient | Data Deficient | Data Deficient |
| | Least Concern | Least Concern | Least Concern | Least Concern |
| | Near Threatened | Endangered | Vulnerable | Near Threatened |
| | Endangered | Critically Endangered | Near Threatened | Vulnerable |
| | Critically Endangered | Vulnerable | Endangered | Endangered |
| | Vulnerable | Near Threatened | Critically Endangered | Critically Endangered |
| | Extinct in the Wild | Extinct in the Wild | Extinct in the Wild | Extinct in the Wild |
| | Extinct | Extinct | Extinct | Extinct |

**10.19i The IUCN Red List Index shows the trends in overall extinction risk for groups of species using a range of values from**

a. 0 to 1

b. 0 to 100

c. 1 to 10

d. 1 to 100

**10.20i Beardsworth and Bryman (2001) have discussed what they have called the 'Disneyization' of zoos. This is a reference to the tendency of zoos to**

a. focus on the keeping of the species of animals that have featured in Disney films

b. construct large, expensive, themed exhibits

c. keep only large attractive animals

d. keep only those animals that children prefer to see

# Advanced

**10.1a** **The Convention on International Trade in Endangered Species of Wild Fauna and Flora 1973 regulates international trade in species listed in**

    a.  two appendices

    b.  three appendices

    c.  four appendices

    d.  five appendices

**10.2a** **Which of the organisations listed below has the following mission statement?**

*To create, deliver, provide training in, and sustain accessible software tools and associated scientific innovations that facilitate science-based decision making in species conservation.*

    a.  Chicago Zoological Society

    b.  Species360

    c.  Species Conservation Toolkit Initiative

    d.  International Union for the Conservation of Nature

**10.3a** **Which of the following could not be the subject of a TAG?**

    a.  A family

    b.  An order

    c.  A class

    d.  All of the above could be the subject of a TAG

**10.4a** **The head office of the European Association of Zoos and Aquaria is located in**

    a.  ZSL London Zoo

    b.  Berlin Zoo

    c.  Barcelona Zoo

    d.  Artis Zoo, Amsterdam

**10.5a** The International Association of Directors of Zoological Gardens was the original name of

    a. EAZA

    b. WAZA

    c. IUCN

    d. BIAZA

**10.6a** The acronym for the system within ZIMS that keeps records of the animals kept by zoos is

    a. KEEPER

    b. RECORD

    c. ARKS

    d. ZOO

**10.7a** As a general rule, zoos and aquariums must comply with

    a. zoo licensing legislation only

    b. zoo licensing legislation and all other applicable legislation including that relating to animal welfare, health and safety, the disposal of hazardous waste and the veterinary treatment of animals

    c. zoo licensing and other legislation relating to animal welfare, and the veterinary treatment of animals only

    d. zoo licensing and other legislation including that relating to animal welfare, health and safety, and the veterinary treatment of animals only

**10.8a** Table 10.3 lists the mission statements published by a major zoo in the United Kingdom over a period of about 25 years. In what chronological order were they used (earliest first)?

    a. 3, 1, 5, 4, 6, 2

    b. 1, 5, 3, 6, 4, 2

    c. 4, 1, 5, 2, 6, 3

    d. 3, 6, 4, 1, 5, 2

**Table 10.3**

| | Mission statement: *Our mission is...* |
|---|---|
| 1 | *...to support and promote conservation by breeding rare and endangered animals, by excellent animal welfare, high quality public service, recreation, education and science* |
| 2 | *...preventing extinction* |
| 3 | *...to promote conservation by breeding of rare and endangered animals and by educational, recreational and scientific activities* |
| 4 | *...to be a major force in conserving biodiversity worldwide* |
| 5 | *...to support and promote conservation by breeding threatened species, by excellent animal welfare, high quality public service, recreation, education and science* |
| 6 | *...to be a major force in conserving the living world* |

**10.9a** **One of the targets set under the United Nations Convention on Biological Diversity 1992 (Target 12) required Contracting Parties to prevent the extinction of known threatened species and improve their conservation status by 2020. This was one of 20**

    a. Cartagena Biodiversity Targets

    b. Nagoya Biodiversity Targets

    c. Washington Biodiversity Targets

    d. Aichi Biodiversity Targets

**10.10a The International Species Inventory System was founded by**

    a. Dr Ulysses Seal

    b. Dr Robert Lacy

    c. Dr Betsy Dresser

    d. Dr William Conway

**10.11a A Regional Collection Plan (RCP) is made by**

    a. an SSP

    b. an EEP

    c. a TAG

    d. a zoo

**10.12a** Article 9 of the United Nations Convention on Biological Diversity 1992 states that:

*Each Contracting Party shall, as far as possible and as appropriate, predominantly for the purpose of comple-menting* in-situ *measures:*

*(a) Adopt measures for the* ex-situ *conservation of components of biological diversity,*

    a.  *from anywhere in the world*

    b.  *preferably in the country of origin of such components*

    c.  *by establishing zoos, aquariums and botanical gardens*

    d.  *including the release of captive-bred animals to the wild*

**10.13a** The Central Zoo Authority is responsible for overseeing the operation of zoos in

    a.  Australia

    b.  the United Kingdom

    c.  the European Union

    d.  India

**10.14a** ISIS was established in

    a.  1965 by 23 zoos in the United States

    b.  1974 by 55 zoos in the United States and Europe

    c.  1982 by 103 zoos in the North America, Europe and Australia

    d.  1995 by 132 zoos across the world

**10.15a** The name of the forerunner of the system within ZIMS that keeps veterinary records of the animals held by zoos was called

    a.  *MedARKS*

    b.  *VetARKS*

    c.  *MedZOO*

    d.  *ZooVET*

**10.16a Which of the following is not part of the ZIMS suite of software?**

    a. ZIMS for Education

    b. ZIMS for Husbandry

    c. ZIMS for Aquatics

    d. ZIMS for *Ex-situ* Conservation

**10.17a The organisation responsible for issuing import and export permits and certificates for specimens of species listed in the Appendices of CITES is**

    a. the CITES Management Authority of a Contracting Party to CITES

    b. the zoo or aquarium wishing to import or export the specimens

    c. the regional zoo organisation of which the zoo or aquarium wishing to import or export the specimens is a member

    d. the International Union for the Conservation of Nature

**10.18a The Laboratory for the Conservation of Endangered Species (LaCONES) is located in**

    a. United States

    b. China

    c. India

    d. Germany

**10.19a The Zoological Information Management System (ZIMS) was first rolled out as a web-based system to replace other zoo record-keeping systems in**

    a. 1998

    b. 2005

    c. 2011

    D. 2016

**10.20a By 2018 Species360 was operating ZIMS in which of the following languages?**

    a. English only

    b. English and Spanish

    c. English, Russian and Japanese

    d. English, Spanish, Russian and Japanese

# 11 Answers

A multiple choice question has a stem (the 'question'), a key (the 'answer') and a number of distracters (wrong answers intended to distract the student from the key). This part of the book contains the key to each question along with a brief explanation of why this is correct and, in some cases, what the distracters mean.

## Chapter 1   History of Zoos and Aquariums

| 1.1f | B | The remains of a variety of animal species have been found at a site in Hierakonpolis, in Upper Egypt, including elephants, hippopotamuses, baboons and hartebeest. This large menagerie dates to around 3500BCE. |
|------|---|---|
| 1.2f | D | This term originated in France. The French word 'ménagerie' is derived from ménager, 'to keep house'. Housekeeping originally included caring for domestic animals. |
| 1.3f | C | 'Zoo' is an abbreviation for zoological gardens. |
| 1.4f | A | The first record of the use of the word 'zoo' was in 1847 with reference to the Clifton Zoological Gardens, now known as Bristol Zoo. |
| 1.5f | D | There was an elephant kept in the Tower of London menagerie from 1255, long before Hagenbeck was keeping and trading in animals. |
| 1.6f | A | Tiergarten Schönbrunn (Schönbrunn Zoo) in Vienna predates the zoos in the distracters and was founded in 1752 by Emperor Franz I, originally as a private menagerie for the imperial family and their guests. It was opened to the public in 1778. |
| 1.7f | C | Tierpark Hagenbeck in Stellingen, Hamburg, Germany opened in 1907 and was the first to use open enclosures with hidden barriers. |

© Paul A. Rees 2021. *Key Questions in Zoo and Aquarium Studies:*
*A Study and Revision Guide* (P.A. Rees)
DOI: 10.1079/9781789249002.0011

| 1.8f | B | Heini Hediger was a Swiss zoologist who was at one time the Director of Zurich Zoo in Switzerland. |
|------|---|---|
| 1.9f | A | *Knut* was a polar bear (*Ursus maritimus*) born in Berlin Zoo, Germany. Many scientists and zoo professionals believe that a zoo environment cannot provide good welfare for polar bears. |
| 1.10f | C | *Jumbo* was captured in the Sudan and sold to the Jardin des Plantes in Paris. From Paris he was transferred to London Zoo in England and later controversially sold to the circus owner and showman Phineas T. Barnum in the United States. |
| 1.11f | D | The dodo (*Raphus cucullatus*) is the emblem of the Durrell Wildlife Conservation Trust based at *Durrell* (Jersey Zoo) in the Channel Islands. The photograph shows a sculpture of a dodo at the entrance to the zoo. |
| 1.12f | D | Each of the individuals owned their own private zoo. |
| 1.13f | B | The Arizona-Sonora Desert Museum is a zoo in the United States that keeps desert animals and plants representative of the local natural history. |
| 1.14f | A | These men travelled around England in the 18th and 19th centuries displaying exotic animals to the public in their menageries. |
| 1.15f | C | *Morgan* was a female killer whale. |
| 1.16f | B | Melbourne Zoo is the oldest zoo in Australia. It was founded in 1857 and opened at its current site in 1862. |
| 1.17f | D | Wild animals were slaughtered in large numbers for public entertainment in ancient Rome. |
| 1.18f | A | The Mappin Terraces are a simulated mountain landscape at London Zoo. |
| 1.19f | B | Giant pandas were sent by China to other countries as an act of diplomacy especially during the Cold War era. |
| 1.20f | A | The WWT opened its first reserve and established its headquarters at Slimbridge in Gloucestershire, England. |
| 1.1i | B | A Pachyderm House would traditionally have kept elephants, rhinoceroses and hippopotamuses; animals with unusually thick skin. |
| 1.2i | B | Dudley Zoo contains the world's largest collection of buildings designed by the Tecton group led by Berthold Lubetkin. These buildings are protected by law because of their historical importance. |
| 1.3i | C | Sir Peter Scott is correct. In addition to being one of the founders of the WWF Scott also founded the Wildfowl Trust (now the Wildfowl and Wetlands Trust). The distracters are other famous naturalists who had a particular interest in ornithology. |

| 1.4i | D | In China a 'Garden of Intelligence' was a menagerie created by Emperor Wen Wang circa 1000BCE. |
|---|---|---|
| 1.5i | A | Frédéric Cuvier was at one time responsible for the Ménagerie du Jardin des Plantes in Paris. The distracters are historically important menageries in Paris and Austria. |
| 1.6i | C | This is a bear pit in Berne where bears were exhibited to the public. |
| 1.7i | A | This book was written by Heini Hediger and was the first book written about zoo biology as a distinct branch of zoology. |
| 1.8i | B | The first SeaWorld was opened in Mission Bay, San Diego, California, in 1964. |
| 1.9i | B | The Exeter Exchange was a building in London that housed a menagerie. |
| 1.10i | D | Philadelphia Zoo is generally considered to have been the first true zoo established in the United States. It opened in 1874. |
| 1.11i | A | Lorenz was one of the founders of ethology. Hagenbeck was an animal dealer and established a zoo in Hamburg; Burton was an architect who designed a number of zoo buildings, especially in London Zoo; Raffles was one of the founders of the Zoological Society of London and its zoo in Regent's Park. |
| 1.12i | C | C is correct. The distracters are the same names matched with the incorrect descriptions. |
| 1.13i | B | The names are those of zoo benefactors. This practice is common in zoos in the United States. |
| 1.14i | A | Henry I established a zoo in the Tower of London where among the animals kept were lions, a polar bear and an elephant. |
| 1.15i | B | This is a sculpture of Gerald Durrell, who established a zoo now known as *Durrell* on the island of Jersey in the Channel Islands. |
| 1.16i | C | This is the entrance to Artis Zoo in Amsterdam. |
| 1.17i | D | Aristotle wrote *History of Animals*. |
| 1.18i | B | The Raven's Cage was constructed in 1828. The photograph does not show the original structure as it was renovated in 1927 and reconstructed in 1948 after sustaining bomb damage. It was originally designed to house macaws. |
| 1.19i | A | Montezuma was an Aztec emperor who kept a large menagerie at Tenochtitlán in what is now Mexico City. |
| 1.20i | D | Henry I created a menagerie in the Royal Park of Woodstock that was later moved to the Tower of London. |
| 1.1a | A | This is a paradeisos. This is the Greek name for such a place derived from the Proto-Iranian word parādayjah (from *pari-* 'around' and *dáyjah* 'wall'). |

| 1.2a | D | This is Carl Hagenbeck. The distracters are the names of other individuals who are associated with zoos and animal exhibits. William Temple Hornaday was the first director of the New York Zoological Park (Bronx Zoo); Phineas T. Barnum was a circus owner and showman famous for purchasing the elephant *Jumbo* from London Zoo; Marlin Perkins was made director of Lincoln Park Zoo in 1944. |
|------|---|---|
| 1.3a | B | The first Safari Park to open in the United Kingdom was at Longleat House in Wiltshire in 1966. Knowsley Safari Park – now Knowsley Safari – opened in 1971. |
| 1.4a | B | This building was designed by Sir Hugh Casson and is perhaps most famous for the fact that it was completely unsuitable for either rhinoceroses or elephants. It no longer houses either. |
| 1.5a | C | George Mottershead founded Chester Zoo not Colchester Zoo. |
| 1.6a | D | Bartlett Society members are primarily interested in zoo history. It is named after Abraham Dee Bartlett (1812-1897), a taxidermist, expert on captive animals and the superintendent of London Zoo. |
| 1.7a | A | London Zoo opened the world's first insect house. |
| 1.8a | D | Ota Benga was a Mbuti man who was exhibited at the Louisiana Purchase Exposition in 1904 and later kept as an exhibit at the Bronx Zoo in New York. He was from one of the indigenous groups of people from the Congo region of Africa previously referred to as pygmies. |
| 1.9a | C | William Temple Hornaday was the first director of the New York Zoological Park, now known as the Bronx Zoo. |
| 1.10a | B | The last captive quagga (*Equus quagga quagga*) died in 1883 at Artis Magistra Zoo, Amsterdam. |
| 1.11a | C | C is correct. The distracters are other Egyptian rulers. |
| 1.12a | C | The Shedd Aquarium is located in Chicago in the Unites States and is named after John Groves Shedd, a wealthy businessman. |
| 1.13a | B | Charles Darwin was a regular visitor to London Zoo. The distracters had long associations with London Zoo. |
| 1.14a | C | Jeanne Villepreux-Power was a marine biologist and invented the first aquarium; Anna Thynne was a marine zoologist who built the first marine aquarium; and Joan Beauchamp Procter was curator of herpetology at London Zoo. |
| 1.15a | D | The image shows Louis XIV's menagerie at Versailles, near Paris, France. |
| 1.16a | A | A is correct. The zoo successfully introduced a number of species into Australia including blackbirds, quail, salmon, camels and sheep. |

| 1.17a | C | This aviary is in ZSL London Zoo. It was conceived by Lord Snowdon and is protected as a listed building. It opened in 1965 and was the first walk-through aviary in Britain. |
| 1.18a | C | C is correct. The distracters are the same book titles matched to the incorrect authors. |
| 1.19a | B | Betsy Dresser is famous for her contributions to developments in assisted reproductive technologies for endangered species. |
| 1.20a | D | D is correct. The distracters are the incorrect zoos matched with the same extinct species of animals. |

# Chapter 2  Zoo and Exhibit Design

| 2.1f | B | Containment is correct. The distracters are words with a similar meaning. |
| 2.2f | D | A reinforced pipe barrier is suitable for retaining large powerful animals. |
| 2.3f | B | Electrified wires capable of giving a mild electric shock are called 'hot wires'. |
| 2.4f | A | Hard standing is a hard surface (usually concrete) provided for animals (especially ungulates) so they have somewhere stable and dry to stand when their paddock is wet. |
| 2.5f | C | Electric fences should not be used as a primary containment barrier. They act as a deterrent once animals experience a mild shock. However they could easily be trampled or damaged by a large animal. |
| 2.6f | D | Glass may be invisible to wild birds and they may be injured or killed if they fly into it. Many birds instinctively avoid birds of prey when they recognise their basic outline. |
| 2.7f | B | The use of a dry moat (i.e. one that does not contain water) is an outdated containment method for some large mammals, especially elephants. These structures are unattractive and pose a risk to the animals as they can fall – or be pushed – into the moat. |
| 2.8f | C | A zoonosis is a disease that is transmissible between humans and animals. Primates, especially great apes, are theoretically susceptible to catching airborne diseases from visitors. |
| 2.9f | A | Only the Zoomesh should interrupt the view of animals. All of the other barriers should be located below the line of sight. |
| 2.10f | A | The overhanging structure is called the 'return'. |

| 2.11f | D | These are psychological barriers. They mark a boundary for the visitors but do not physically prevent them from entering the area intended to be exclusively occupied by the animals. |
| 2.12f | D | Naked mole rats live in networks of underground tunnels and chambers. Zoos often exhibit them in systems of transparent plastic tubes in a room with little lighting. |
| 2.13f | B | Pinioning is a technique that prevents birds from flying. It is widely used for some flying birds – such as ducks, geese and flamingos – so that they can be exhibited in open pens or allowed to wander around the public areas of a zoo. Pinioning involves the removal of the joint of the bird's wing farthest from the body (pinion joint) on one wing only. |
| 2.14f | D | Okapi are shy forest animals and are inappropriate for a savannah exhibit. |
| 2.15f | B | This is a biofloor and is widely used in modern primate exhibits. |
| 2.16f | C | It is possible to keep several lemur species together such as ring-tailed lemurs (*Lemur catta*), black lemurs (*Eulemur macaco*) and black-and-white ruffed lemurs (*Varecia variegata*). |
| 2.17f | A | The term 'herps' refers to reptiles and amphibians. In a temperate climate these animals need to be kept in an environment where the temperature and humidity can be carefully controlled. This normally means housing them in an indoor terrarium. |
| 2.18f | C | A guillotine door consists of a vertical plate of metal that slides up and down in a metal frame. To open the door it must be pulled up, usually with a steel cable operated through a pulley system. |
| 2.19f | D | Of the species listed, only the golden lion tamarin is active during daylight hours. The other species are nocturnal and usually exhibited in a small mammal house or nocturnal house where the lighting schedule is reversed so it is dark during the day and visitors are able to see the animals when they are active. |
| 2.20f | B | This is an immersion exhibit because visitors are able to enter the world of the gorillas and see them at close quarters. The floor of the gorilla house is at the same level as the ground in the enclosure and, although there is a glass barrier separating them, visitors feel as if they are sharing the same landscape with the gorillas. |
| 2.1i | C | Lubetkin and his associates in the Tecton Group designed a number of modernist enclosures in the United Kingdom, predominantly from concrete. Many of these structures still exist and are protected because of their historical and architectural importance, especially at London Zoo and Dudley Zoo in the West Midlands. |

| 2.2i | B | These structures are collectively called 'furniture'. |
|---|---|---|
| 2.3i | A | Jon Coe is a designer who studied landscape architecture at Harvard University and has specialised in zoo design since the mid-1970s. |
| 2.4i | D | Exhibit design has evolved from cages and pits (1st generation) to open enclosures with 'invisible' barriers (2nd generation) and then to naturalistic immersive exhibits in which the animals and visitors share the same landscape (but not the same space). |
| 2.5i | C | This is a rotational exhibit. It extends the total area used by each species on a time-share basis and exposes each species to the scents of the others. |
| 2.6i | A | Both doors in a double door entry system should never be open at the same time. |
| 2.7i | D | The purpose of the stand-off barrier is to keep visitors away from the primary containment barrier (in this case a chain-link fence). The space between the stand-off barrier and the main fence is often filled with vegetation to prevent visitors from reaching the fence. |
| 2.8i | B | A transfer chute is a secure area through which animals are moved to get from one location to another, for example, between different cages within their indoor accommodation or from their indoor accommodation to their outdoor enclosure. |
| 2.9i | D | A thermal gradient should be maintained so that animals can select their preferred temperature. This is often achieved by placing a heat source (often an infrared lamp) at one end of the vivarium. |
| 2.10i | C | Invertebrates are under-represented in the zoo community as a whole even though most of the animal species on Earth are invertebrates. |
| 2.11i | A | Lancelot 'Capability' Brown used this technique to create impressive vistas across the English countryside. |
| 2.12i | B | The first naturalistic enclosure in the world is believed to have been the gorilla exhibit opened at Woodland Park Zoo in 1978. |
| 2.13i | C | C is correct. Both hot wires should be on the inside of the fence: one at the bottom to keep the animals away from the fence, and one at the top to deter them from attempting to climb over the top of the return if they were to avoid the lower hot wire. |
| 2.14i | C | C is located away from the main route and is the best location. Most visitors tend to avoid straying away from the main routes around a zoo so this location should be visited by fewer members of the public than the other locations. A is a particularly poor location because it is near the entrance and so exposed to visitors on their way into the zoo and on their way out. |

| 2.15i | A | This term was first coined in the mid-1970s. Landscape immersion might be achieved by lining a path to an exhibit with tall bamboo plants so that the visitor experiences the illusion of coming across an animal in a clearing where there is a gap in the vegetation. |
|---|---|---|
| 2.16i | B | Lubetkin's enclosures were predominantly made of concrete. |
| 2.17i | D | This is the giraffe house at London Zoo. |
| 2.18i | C | A circadian rhythm is one that has a periodicity of approximately 24 hours and is determined by light levels. Seasonal changes in behaviour may be reversed in species that have adapted to the northern hemisphere and are kept in captivity in the southern hemisphere, and vice versa. |
| 2.19i | D | A post-occupancy evaluation of an exhibit takes place after the animals have been added and visitors have been allowed access. It may be used to provide the zoo with information about its effectiveness and inform future designs. |
| 2.20i | A | A puparium is a place where butterfly and moth cocoons (pupae) are kept while the insects undergo metamorphosis into their adult form. |
| 2.1a | B | The Penguin Pool is one of several enclosures at London Zoo designed by Lubetkin. |
| 2.2a | C | Zoo360 is located in Philadelphia Zoo and allows animals to explore sections of the zoo through mesh tunnels providing animals and visitors with novel viewpoints. |
| 2.3a | D | These are the equivalent of safari parks. The term 'open range zoo' is used in Australia to apply to, for example, Werribee Open Range Zoo and Taronga Western Plains Zoo. |
| 2.4a | B | *Gunite* is a material used to created simulated rock surfaces in animal enclosures and elsewhere in zoos. |
| 2.5a | D | The Buffalo House was built to house American bison (*Bison bison*) and was the first building constructed at the Smithsonian's National Zoological Park, Washington, D.C. |
| 2.6a | A | A paludarium is a type of vivarium that simulates a swamp or rainforest habitat and contains underwater areas. |
| 2.7a | A | The 'vertical zoo' is a design concept for an urban zoo. The distracters are fictitious. |
| 2.8a | C | During the 'Disinfectant Era' animal cages were constructed with tiled walls and other features that made them very easy to clean. |
| 2.9a | D | During the 'Disinfectant Era' the welfare of animals declined and the visitor experience was poor because the animals were exhibited in sterile environments. |
| 2.10a | D | N is the only enclosure where it is possible for an animal to position itself so that it is more than 7m from the perimeter. |

| | | |
|---|---|---|
| 2.11a | B | Land area = (1.67m² x 6 birds) + (0.84m² x 8 birds) = 16.74m²; Water volume = water surface x 1.33m = 16.74 x 1.33m = 22.26m³. |
| 2.12a | C | The sons of Carl Hagenbeck, Heinrich Hagenbeck and Lorenz Hagenbeck, were granted this patent for enclosure designs where the barriers were 'invisible'. |
| 2.13a | A | This concept for a new type of zoo reverses the usual situation and confines the public in small areas and makes large spaces available to the animals. |
| 2.14a | C | Enclosure designs should avoid creating areas from which subordinate animals cannot escape when cornered by an aggressive dominant animal. Enclosure C has rounded corners. Each of the other designs has at least four right-angled corners where individuals could easily be trapped by others. |
| 2.15a | D | This is the ground skirt. The distracters are fictitious in this context. |
| 2.16a | D | Bear pits are designed so that visitors look down on the bears from the periphery of the pit. Putting humans in an elevated position in relation to the animals could be seen as encouraging a dominionistic attitude towards them because the people are literally looking down on the animals. |
| 2.17a | B | B is correct. All of these are themed exhibits. The distracters are the same zoos matched to the incorrect exhibits. |
| 2.18a | B | In protected contact the keepers do not share the same space as the elephants but work with them from behind a metal fence. This includes hinged doors that give access to different parts of the elephant to allow the body to be inspected and for veterinary treatment. |
| 2.19a | A | Cultural resonance refers to the development of a mutual understanding of situations seen from a cultural perspective. The lives and survival of indigenous peoples are inextricably linked to their environment and they depend upon wild animals and plants to provide them with food and materials in a way that is alien to people living in industrialised parts of the world. In addition, human-animal conflict is a significant economic and conservation issue in many parts of the world. Some zoos attempt to show the links between human cultures and wildlife by, for example, constructing facsimile buildings indicating the presence of humans in, or very closely associated with, the animals' habitats. |
| 2.20a | C | The temperature range to which a species is exposed in the wild depends upon the latitudinal extent of its geographical range and the elevation at which it lives because temperature falls with altitude. Species that live in tropical areas are exposed to low temperatures if they live at high altitude. |

# Chapter 3   Aquariums and Aquatic Exhibits

| 3.1f | C | The Sumerians kept wild-caught fish in ponds for food in southern Mesopotamia. |
|---|---|---|
| 3.2f | A | The Fish House was opened in London Zoo in 1853. |
| 3.3f | D | The Two Oceans Aquarium is in Cape Town, South Africa. |
| 3.4f | B | Ozone is used to disinfect water. It is a powerful oxidising agent and will kill bacteria and viruses. |
| 3.5f | A | The accumulation of nutrients in water resulting from contamination with animal faeces will cause eutrophication. This nutrient enrichment will cause an increase in algae, depletion of fish stocks and a reduction in water quality. |
| 3.6f | B | A basic aquatic life support system removes unwanted materials from water by mechanical filtration before it undergoes biological filtration and then chemical filtration. |
| 3.7f | B | The standard salinity of natural seawater is 35 parts per thousand. |
| 3.8f | C | Any pH greater than 7.0 is alkaline so tanks 3 and 4 contain alkaline water. |
| 3.9f | D | This was opened by the showman and circus owner Phineas T. Barnum. The distracters are the names of people associated with zoos in the United States. |
| 3.10f | B | The first oceanarium was opened in Florida in 1938. |
| 3.11f | A | A hydrometer can be used to measure the specific gravity of water. A hygrometer measures humidity. |
| 3.12f | D | The bed could be made of sand, silica chips or small plastic pellets which provide a large surface area for the bacteria needed for the biological treatment of the water. |
| 3.13f | D | The term 'cycling time' refers to the time it takes for a new aquarium to build up sufficient nitrifying bacteria to convert ammonia and nitrite to nitrate efficiently. Once ammonia and nitrite levels reach zero the aquarium has completed its first cycle and it is safe to add more fish. |
| 3.14f | B | Optimum water temperature for most tropical fish species is 22-28°C (72-82°F). |
| 3.15f | C | 150cm x 50cm x 80cm = 600,000 cm$^3$ or 600 litres. This would weigh 600kg. |
| 3.16f | A | A 'pig' is a ball that can be forced through a water pipe to clear away accumulated biological growth. This is likely to be a particular problem for a marine aquarium that takes water from the sea. |

| 3.17f | D | A refugium could have all of the functions listed and provides a safe place for organisms to be maintained that would not survive in the main tank. It is connected to the main tank and shares the same water supply. |
|-------|---|---|
| 3.18f | B | A refractometer may be used to measure salinity. |
| 3.19f | D | This process is called maturation and the time it takes in a newly established aquarium is the cycling time. |
| 3.20f | C | Polymethyl methacrylate (acrylic glass) is a synthetic resin used to make windows and transparent tunnels in aquariums. |
| 3.1i | A | A calcium reactor is a device that supplies calcium to the water in a marine or reef aquarium where it is used by corals to construct their skeletons. |
| 3.2i | D | Carbon dioxide dissolved in water creates carbonic acid and would lower the pH (i.e. make it more acidic). |
| 3.3i | B | The reversal of water flow through a filter is known as backwashing. |
| 3.4i | D | The rotating drum in a rotary vacuum filter pulls solids out of the water which are then removed as they pass a knife blade. |
| 3.5i | A | A baffle is a structure that alters the water flow in an aquarium system. |
| 3.6i | C | These are bioballs and they provide a large surface area for the biofilm in some types of biological filter. |
| 3.7i | B | A foam fractionator is also called a protein skimmer. A calcium reactor is a device that adds calcium to the water. Distracters C and D are fictitious in this context. |
| 3.8i | A | Open aquarium systems obtain their water directly from the sea. This water contains living organisms that will eventually accumulate on the inside of the pipes and need to be removed. This accumulation of biological material is called biofouling. |
| 3.9i | D | Turbidity is affected by the presence of particulate matter, dissolved gases and chemicals in aquarium water. |
| 3.10i | D | Specific gravity is dimensionless so has no units. |
| 3.11i | C | UV light is used to kill microorganisms. |
| 3.12i | B | The figure shows a fluidised bed filter. In this type of filter the medium is held in suspension by a flow of pumped water thereby making the entire surface of the medium available for bacterial growth. |
| 3.13i | A | Redox potential measures water quality. Redox potential is a measure of the propensity of a chemical to gain or lose electrons through ionisation, thereby becoming either reduced or oxidised, respectively. The lifespan of bacteria in water is determined by its redox potential. Many important biochemical reactions are oxidation or reduction reactions (e.g. ammonia → nitrite → nitrate). The redox potential will determine which reactions are prevalent. |

| 3.14i | D | Water hardness is largely the result of the presence of calcium and magnesium compounds. |
|---|---|---|
| 3.15i | B | Two litres of fresh water should be added – not salt water – because the salts remain in the tank. Adding salt water would increase the salinity. |
| 3.16i | C | The specific gravity of marine water kept at 24° – 25°C (75 – 77°F) should be 1.021 – 1.024. |
| 3.17i | B | Seeding water in a new tank – that is one that has not matured – helps to avoid new tank syndrome. The term 'new tank syndrome' refers to the rising ammonia and nitrite levels that occur when fish are added to a new aquarium tank. |
| 3.18i | C | A sump is an accessory tank that holds water outside the main tank (but is connected to it) thereby contributing to the total quantity of water in circulation. |
| 3.19i | D | A sump might contain equipment used to treat the water but would not contain fish or other animals. |
| 3.20i | B | This is usually known as a touch pool. |
| 3.1a | A | Chemicals are removed by adsorption. This is the adhesion of atoms, ions or molecules to a surface, in this case charcoal. |
| 3.2a | D | Appropriate levels of calcium and magnesium in aquarium water and a suitable pH must be maintained for the successful growth of corals. |
| 3.3a | A | In a biological filter nitrifying bacteria convert ammonia to nitrite and then to nitrate. |
| 3.4a | B | The process of reverse osmosis requires the presence of a semi-permeable membrane which separates ions and larger particles from water. A pressure is applied to overcome the osmotic pressure of the aquarium water. |
| 3.5a | D | This water needs to be treated and recirculated (because the system is closed) but salt does not need to be added as this is not lost from the system when water evaporates. |
| 3.6a | C | A sump is always located outside the main tank and increases the amount of water in circulation. |
| 3.7a | B | The amount of oxygen held by water decreases as temperature increases, i.e. there is a negative correlation between oxygen concentration and temperature not a positive correlation. |
| 3.8a | D | Any of the listed events could cause a spike in the ammonia or nitrate levels in a mature tank due to damage to, overloading of, or failure of the biological filtration system. |

| 3.9a | C | A heat exchanger transfers heat from hot water to cold water so the hot water loses heat and the cold water gains heat. Heat exchangers are used in public aquariums to control the water temperature. They may, for example, transfer the heat generated from the public areas (from lighting and other heat sources) to the water used in tanks containing tropical fishes. |
|------|---|---|
| 3.10a | D | The water in a tank designed to hold jellyfish has a slow circular flow to mimic currents in the ocean and keep them off the bottom of the tank. |
| 3.11a | A | An adsorption granulate removes phosphates and silicates by the process of adsorption (i.e. they stick to its surface). |
| 3.12a | B | Ozone acts as a biocide and is a very powerful oxidising agent. |
| 3.13a | A | Redox potential is measured in millivolts. |
| 3.14a | B | A MultiCyclone is a prefiltration device that removes sediment using centrifugal force. It has no moving parts and no filter media. |
| 3.15a | D | Organic compounds bind to small bubbles in a protein skimmer producing a frothy mixture containing solid waste and uneaten food which is collected by a receptacle and removed. |
| 3.16a | B | Philip Henry Gosse first used the term 'aquarium' in the modern sense of the word and invented the institutional aquarium. He created the first successful aquarium for the long-term maintenance of marine organisms and described it in 1854. |
| 3.17a | B | *Tilikum* was a killer whale. |
| 3.18a | C | *Turning the Tide* is the title of WAZA's 2009 global conservation and sustainability strategy for aquariums. The distracters are fictitious in this context. |
| 3.19a | B | Ammonia peaks first. This is converted to nitrite so this produces the second peak. The third peak is the result of nitrite being converted to nitrate. |
| 3.20a | A | The level of M (nitrate) has dropped at point X because some of the water has been changed. |

# Chapter 4   Visitor Studies, Zoo Education and Zoo Research

| 4.1f | D | Biophilia is the emotional affiliation that many people feel towards animals and one of the reasons they enjoy visiting zoos and wild places. |
|------|---|---|
| 4.2f | C | Most visitors to zoos attend as part of a family, especially those with young children. |

| 4.3f | A | Data collected for a single day are most accurate. A visitor who attends a zoo on a particular day is unlikely to leave that zoo and return later in the day, be issued with a second ticket and therefore be counted twice. Data collected for a week, month or year will undoubtedly count all of the admissions to the zoo including multiple visits by the same people. If a zoo claims to have a million visitors a year this total is unlikely to mean a million different visitors. |
|---|---|---|
| 4.4f | D | Some individual animals, and some species, may react positively to visitors (e.g. by approaching them and seeking interaction) and others may act negatively (e.g. by moving away from visitors or acting aggressively). Others may ignore visitors completely. It depends on the species and the individual. |
| 4.5f | B | Orientation is concerned with knowing one's location. In a zoo it would refer to knowing where you are standing in the zoo at a particular point in time: not being 'lost'. Zoos may assist in this by providing marked meeting points, printed paper maps and signage containing a map with an arrow indicating 'You are here'. |
| 4.6f | C | A number of studies have shown that large mammals are the most popular animals with zoo visitors. |
| 4.7f | D | The role of an interpretation board is to provide and explain information; to interpret an exhibit for the visitor. |
| 4.8f | D | Signs pointing to other locations tell a person where those places are but not where the sign is. Signs that use outlines of animals do not assume an understanding of a particular language so are useful for young children and foreign visitors. |
| 4.9f | A | These images are pictograms that communicate to visitors the direction in which they need to walk to find particular types of animals and animal houses. |
| 4.10f | C | Visitor circulation describes the pathways taken by visitors around a zoo. |
| 4.11f | A | *International Zoo News* is not a peer-reviewed journal. |
| 4.12f | B | Most zoo research is conducted on mammals. |
| 4.13f | B | Enclosure A contains the control. Enclosure B contains the experimental group because in this enclosure the method of feeding has been altered. |
| 4.14f | A | The *International Zoo Yearbook* has published data on zoos including the animals they keep and their visitor numbers on a more-or-less annual basis since 1960. |
| 4.15f | C | The time spent at an exhibit is the dwell time. The distracters are fictitious in this context. |

| 4.16f | D | Active animals, those being trained and exhibits where public participation is possible all hold the attention of visitors for longer than animals and exhibits that do not have these characteristics. |
|-------|---|---|
| 4.17f | C | The mean is a measure of central tendency (a middle value) and the standard deviation is a measure of dispersion (variation around the mean). |
| 4.18f | B | Whether or not the data represent a 1:1 ratio of males to females may be tested using a chi-squared test. |
| 4.19f | A | This sign is an example of an infographic: it uses images, a map and minimal text to explain a topic. |
| 4.20f | B | The 'visitor attraction' model assumes that visitors are attracted to active animals. |
| 4.1i | A | Viewing point is correct. A glade is an open area in a forest. |
| 4.2i | D | D is correct. The distracters are the same journals linked to the incorrect organisations. |
| 4.3i | B | Interpretation boards that contain very large quantities of information are likely to be ignored by most visitors. The target audience needs to be identified, e.g. is the board aimed at young children or adults? It should follow the brand guidelines of the zoo and be located in a prominent position. |
| 4.4i | D | This is a longitudinal study as it looked at the same animals over an extended period of time. |
| 4.5i | C | An interpretation board is used to explain an exhibit to the visitor. A zoo map, a zoo guidebook (normally including a map), and signposts assist visitors in finding their way around the zoo. |
| 4.6i | D | The zoo contained 76 exhibits. The family spent 370 minutes in the zoo. They spent 45 minutes eating lunch and 30 minutes on the playground. This left 370-75=295 minutes to look at the animals assuming no time was used up moving between exhibits. The average amount of time spent at each exhibit was 295/76=3.9 minutes. |
| 4.7i | C | CITES is an acronym for the Convention on International Trade in Endangered Species of Wild Fauna and Flora 1973. Zoos sometimes display collections of specimens confiscated at airports and elsewhere. |
| 4.8i | B | The Zoos Directive requires zoos located in the territories of Member States of the European Union to, among other things, have an education function. |
| 4.9i | D | This method of pricing tickets can maximise income (by charging a premium at popular times, e.g. school holidays and weekends in summer); even out visitor attendance thereby reducing the crowded conditions that reduce visitor satisfaction on busy days; even out staffing requirements. |

| 4.10i | B | The advertised times of the keeper talks would require visitors to move around the zoo in an illogical manner and backtrack if they wanted to attend all of the talks. It would be more sensible to set the times of the talks to match the sequence in which most visitors encounter the exhibits. |
|-------|---|---|
| 4.11i | A | Red-necked wallabies are native to Australia and nilgai (bluebuck) are indigenous to the Indian subcontinent. Keeping these animals in the same enclosure sends a confusing educational message as they would never be seen together in the wild. |
| 4.12i | D | The *International Zoo Yearbook* is published by the Zoological Society of London. |
| 4.13i | C | These types of documents are collectively called the 'grey literature'. The distracters are fictitious in this context. |
| 4.14i | B | This was a cross-sectional study because the authors inferred the pattern of development of behaviour in chimpanzees by examining individuals of different ages at a specific point in time. If they had followed the development of behaviour over a period of years in a number of chimpanzees born at the same time this would have a been a longitudinal study. |
| 4.15i | D | Bitgood called this behaviour 'dominant path security'. |
| 4.16i | C | Time spent feeding is a continuous variable and in this graph it is the dependent variable because its value depends upon the time of day (the independent variable). |
| 4.17i | A | This was to ensure that each student would record behaviours in an identical fashion and thereby ensure inter-observer reliability. |
| 4.18i | B | Annual visitor attendance is normally reported as the number of visits made to the zoo rather than the number of individual visitors. The latter would be impossible to establish unless zoos recorded the name of every visitor and removed any duplicate visits in the year. |
| 4.19i | C | Where exhibits contained flagship and non-flagship species visitors spent more time watching the flagship species. |
| 4.20i | B | This phenomenon has been observed in museums and other visitor attractions and is called 'museum fatigue'. |
| 4.1a | D | This is dynamic pricing as it is capable of responding to changes in visitor behaviour with time. |
| 4.2a | B | Intragroup aggression, stereotypic behaviour and autogrooming all increased when visitor density was high. |
| 4.3a | C | This balancing of cost against reward is known as the general value principle. The distracters are fictitious in this context. |

| 4.4a | D | Visitors tend to walk in a straight line, travelling the shortest distance between two points. |
|------|---|---|
| 4.5a | A | Studies using a single animal living in a zoo have been published on an Asian elephant, a chimpanzee, a giraffe and a hippopotamus. |
| 4.6a | B | The most popular method of wayfinding in the zoo was to use a hand-held map. |
| 4.7a | A | This is conceptual orientation. The distracters are fictitious in this context. |
| 4.8a | A | Melton's exit gradient theory is correct. The distracters are fictitious in this context. |
| 4.9a | D | D is correct. The distracters are the same terms matched with the incorrect actions or behavioural changes. |
| 4.10a | C | International Zoo Educators Journal is correct. The other journal titles are fictitious. |
| 4.11a | C | Zoo Biology was first published in 1982. |
| 4.12a | A | This is a Monte Carlo simulation as its results are based on chance (like the outcome of bets in a casino in Monte Carlo). |
| 4.13a | B | A negative correlation coefficient indicates that as the independent variable (temperature) decreases the dependent variable (frequency of stereotypic behaviour) increases. |
| 4.14a | B | The value of $r$ was statistically significant at the 5% level, i.e. this result could have occurred by chance on fewer than 5 in 100 occasions ($P<0.05$). |
| 4.15a | D | A meta-analysis draws together the results of a number of different studies that address the same question. |
| 4.16a | C | This is an example of pseudoreplication because the mean was calculated from 60 samples that were not independent. The number of bears sampled was just 6. |
| 4.17a | C | Although it is possible to determine a relationship between maximum daily temperature and the frequency of dusting behaviour from these data it is not possible to determine cause and effect. It may be that elephants dusted more on warmer (sunnier) days to protect their skin from the sun but they may do this for some other reason. |
| 4.18a | A | This repeated redefining of abnormal behaviour is an example of observer drift. |
| 4.19a | C | The results obtained from scan samples are highly reliable but have low validity. That is, the values vary in a similar pattern to those obtained from the video analysis in a reliable way but the actual values from the scan samples are all too low. |

| 4.20a | D | It is not possible to say from the correlation coefficient value alone whether or not it is statistically significant. Any two points plotted on a graph will always have a correlation coefficient of +1.0 or -1.0 (the highest values possible) because they fall on a straight line. (Note that the + and − sign indicate the direction of the relationship not its strength). However, this relationship is not statistically significant because the sample size ($n=2$) is too small. |
|-------|---|---|

# Chapter 5    Nutrition and Food Presentation

| 5.1f | B | Ruminants collect plant material and swallow it twice. The first time it passes down the oesophagus and enters the rumen then the reticulum. The food is then regurgitated into the buccal cavity and the animal 'chews the cud' (cud being the semi-digested food that has been regurgitated). The food is then swallowed a second time and passes down the oesophagus to the omasum and passes to the abomasum before continuing to the small intestine. |
|-------|---|---|
| 5.2f | C | Elephants are not ruminants and have a simple stomach. |
| 5.3f | C | Liver has a high vitamin A content. |
| 5.4f | D | A honey badger is a member of the weasel family. The distracters are rodents and possess incisors that continue growing throughout life so need to be worn down. |
| 5.5f | B | This is a mineral lick used to supplement mineral intake. |
| 5.6f | C | Orchard-grown fruit is often fed to animals in zoos. It has a higher sugar content and a lower protein and fibre content compared with wild fruits and contributes to dental disease, obesity and diabetes in some primates. |
| 5.7f | B | Cafeteria-style feeding consists of offering animals a wide choice of foods so that they are able to choose what to eat; a complete feed is one that is nutritionally complete, i.e. it has been formulated to meet the animal's requirements for energy, protein, vitamins, minerals, etc.; in *ad libitum* feeding animals are able to access food continuously; *ad hoc* feeding has no meaning in this context. |
| 5.8f | A | Fish-eating animals are piscivores. |
| 5.9f | D | Malnutrition may result from having too little food (or too little of particular foods); nutrient excesses that may result, for example, in obesity; the inability to absorb some nutrients. |
| 5.10f | C | Micronutrients – including iron – are required in relatively small quantities. |
| 5.11f | A | Hardbills have strong bills adapted to breaking open seeds. |

| 5.12f | C | Birds such as pigeons and doves. |
|-------|---|----------------------------------|
| 5.13f | D | Feeds for farm animals have been developed to enable rapid and efficient weight gain, high milk yield in species kept for their milk and high egg production in birds. |
| 5.14f | D | Marine mammals obtain their water from their food and their metabolism. |
| 5.15f | B | The chimpanzees developed rickets due to a lack of vitamin D. This was the result of insufficient exposure to the sunlight required for vitamin D synthesis. |
| 5.16f | A | A folivore feeds on leaves. |
| 5.17f | C | San Diego Zoo grows *Acacia* for its giraffes, *Eucalyptus* for the koalas and bamboo for the red pandas. |
| 5.18f | A | Gustation is the sense of taste. |
| 5.19f | C | The amino acids that cannot be synthesised by an animal's body are called essential amino acids. The distracters are fictitious in this context. |
| 5.20f | C | This is scatter feeding and is used, for example, with chimpanzees, monkeys and elephants. |
| 5.1i | D | These structures are concretions formed when minerals, especially calcium, accumulate around a foreign body. |
| 5.2i | B | Vitamin $B_1$ is thiamine. Some fish species are known to contain high levels of thiaminase: an enzyme that destroys thiamine. |
| 5.3i | A | In the wild lions do not feed every day. To simulate their natural feeding pattern many zoos allow lions to gorge themselves on food on some days and leave them fasting on other days. |
| 5.4i | D | This may involve the consumption of hair, soil, metal or plastic objects, etc. The term is derived from the Latin word for a magpie (*Pica pica*). |
| 5.5i | A | This practice is generally known as gut loading. |
| 5.6i | B | An unintentional gradual change to an animal's diet that occurs over a period of time is known as dietary drift. This may occur, for example, because food is not regularly weighed so the quantity given unintentionally increases with time. |
| 5.7i | C | Overfeeding (resulting in obesity) and underfeeding can suppress reproduction in some species. |
| 5.8i | C | Metabolomics is the measurement of all metabolites in an organism to, for example, diagnose a complex metabolic disease and identify inborn errors of metabolism. This allows the development of new therapeutic treatments and the discovery of new biomarkers. |
| 5.9i | A | One form of vitamin E is called α-tocopherol. |

| 5.10i | D | All of the companies using the listed brands manufacture exotic animal foods. |
|---|---|---|
| 5.11i | B | Dietary intake is calculated as the weight of the food offered to the animal minus the weight of the food it does not eat. |
| 5.12i | B | In cafeteria-style feeding animals are allowed to select from a range of different foods. However, given a choice, individuals often choose food items that do not comprise a balanced diet. |
| 5.13i | D | Colostrum contains immunoglobulins (antibodies) that help to protect young animals from infections. |
| 5.14i | C | Model species is correct. For example, the domestic dog is used as a model for large canids when determining their nutritional requirements. |
| 5.15i | A | Equids, rhinoceroses and tapirs are hindgut fermenters; bison are ruminants. |
| 5.16i | C | This is hypervitaminosis D. Hypovitaminosis refers to a vitamin deficiency; avitaminosis is a group of diseases caused by one or more vitamin deficiencies; multivitaminosis is fictitious in this context. |
| 5.17i | C | When an animal is fed a maintenance ration it is given its daily feed allowance in the correct quantity with the components in the correct proportions with the intention of maintaining life indefinitely without any change in weight. |
| 5.18i | D | The basal metabolic rate is the quantity of energy per unit time required to keep the body functioning at rest. |
| 5.19i | A | A is incorrect. Storage conditions can affect the quality and thus the palatability of food. |
| 5.20i | C | Calcium is especially important in egg shell production as the shell is made largely of calcium carbonate. |
| 5.1a | D | A lizard may stop feeding for any of these reasons and for many others. |
| 5.2a | B | Chitin is a structural polysaccharide found in the exoskeleton of arthropods and fungal cell walls. |
| 5.3a | D | All of the effects listed occurred in rhesus macaques whose calorific intake was restricted. |
| 5.4a | A | Haemosiderosis is also known as iron storage disease: an accumulation of iron in the body. |
| 5.5a | D | Homiotherms need more energy than poikilotherms of the same size (not less) because they need to generate heat to maintain their body temperature. |

| 5.6a | C | A faecal condition scoring system does not use the results of chemical analyses of the faeces to derive a score. |
|---|---|---|
| 5.7a | C | Hyponatraemia is sodium deficiency and pinnipeds suffering from this require dietary salt supplements. |
| 5.8a | A | Sirenians are manatees (sea cows) and are herbivores. Cetaceans are dolphins, whales and their relatives; phocids are true ('earless') seals; odobenids are walruses. |
| 5.9a | A | Polar bears have a high dietary requirement for vitamin A. In the wild they obtain this from the livers of their prey. |
| 5.10a | D | Anabolism is the synthesis of complex molecules from simpler molecules in living things; glycolysis is an anaerobic metabolic pathway that coverts glucose to lactic acid in animal cells. |
| 5.11a | B | The dry matter digestibility of food may be determined by the AIA marker technique. This is an acceptable natural marker for determining dry matter digestibility. |
| 5.12a | D | All of the listed options may cause a vitamin deficiency. |
| 5.13a | A | Pinnipeds (seals) produce little lactase: the enzyme that digests the lactose in milk. |
| 5.14a | A | *Zootrition* and *Fauna* are both computer packages used by zoos to manage animal diets. |
| 5.15a | D | The mass-specific metabolic rate of an animal is measured as the number of litres of oxygen used per kilogram of body weight per hour. |
| 5.16a | C | Nutritional wisdom and euphagia mean the same thing: the capacity of an animal to select foods that meet its nutritional needs and avoid harmful foods. |
| 5.17a | B | This is the formula for assimilation efficiency. |
| 5.18a | C | Aversive learning is correct. The distracters are fictitious in this context. |
| 5.19a | B | Aspergillosis is caused by a bacterium. |
| 5.20a | D | Each of the options can have a negative effect on the quality of the food or deny the animals an enrichment opportunity. |

# Chapter 6    Reproductive Biology and Genetics

| 6.1f | A | All of the genes in a population of organisms are its gene pool. The distracters are fictitious in this context. |
|---|---|---|
| 6.2f | D | Panda cubs are born at a very early stage of development. |

| 6.3f | D | A white tiger is a naturally occurring variant of *Panthera tigris* that has two copies of a recessive gene that cause white coat colour. A wholpin is a hybrid between a female bottlenose dolphin (*Tursiops truncatus*) and a male false killer whale (*Pseudorca crassidens*); a zonkey is a hybrid between a donkey and a zebra; a liger is a hybrid between a male lion (*Panthera leo*) and a female tiger (*P. tigris*). |
|---|---|---|
| 6.4f | C | A cloaca is present in birds, reptiles, amphibians and monotremes (egg-laying mammals). |
| 6.5f | C | The optimum time for a female mammal to be inseminated varies with the species. In some species ovulation only occurs after the female has been stimulated by copulation. |
| 6.6f | B | An ectopic pregnancy occurs when a fertilised egg (zygote) implants somewhere other than in the uterus, e.g. in the Fallopian tube. |
| 6.7f | A | The term 'fecundity' refers to the potential of an individual for reproduction. Fecundity takes into account birth rate and offspring survival. |
| 6.8f | D | A tigon is a hybrid formed from a cross between a male tiger and a female lion. A cross between a male lion and a female tiger is a liger. |
| 6.9f | A | Luteinising hormone (LH) is produced by the anterior pituitary gland and a surge in LH triggers ovulation. |
| 6.10f | C | Oxytocin is a hormone that, among other things, causes uterine contractions during labour. |
| 6.11f | D | Oviparous animals produce eggs from which the young emerge after they have been laid. |
| 6.12f | B | Some zoos breed animals and cull the unwanted offspring. This produces young animals that attract visitors and also allows animals to continue to reproduce and rear young rather than receive contraception that could result in infertility. |
| 6.13f | C | In species of birds in which males and females incubate the eggs both species possess a brood patch: a featherless area of skin that transfers body heat to the eggs. |
| 6.14f | D | Testiconid mammals such as elephants and dolphins have internal testicles. |
| 6.15f | C | This organ is the vomeronasal organ or Jacobson's organ. |
| 6.16f | A | A is correct. The distracters are the same terms matched with the incorrect definitions. |
| 6.17f | B | Sex organs are primary sexual characteristics. |

| 6.18f | C | The sex of the offspring of turtles, alligators and frogs may be determined by the temperature at which their eggs are incubated. This phenomenon is not observed in most birds but is known to occur in some. |
|---|---|---|
| 6.19f | B | B is correct. The distracters are the same names for offspring matched with the incorrect species. |
| 6.20f | D | Those animals that are viviparous give birth to miniature adults. This is the mode of reproduction used by most mammals. |
| 6.1i | A | This is a survivorship curve and is usually based on the fate of 1000 animals either born in the same year or the ages of those present in a population at the same time. |
| 6.2i | C | All of the chromosomes that are not sex chromosomes are called autosomes. |
| 6.3i | B | Genetic drift is the loss of genes as a result of random processes and is particularly important in influencing the gene pools of small populations. |
| 6.4i | B | $$N_e = \frac{4(76 \times 98)}{(76 + 98)} = 171$$ |
| 6.5i | A | A dynamic life table is a mortality schedule for a group of animals that were all born at the same time (e.g. in the same breeding season). The individuals are followed throughout their lives until death and the death rate between consecutive age groups is calculated. This is continued until the last animal in the cohort has died. |
| 6.6i | C | This is the founder effect and may be a problem in captive breeding programmes. The use of a small number of founding animals will result in inbreeding, especially if they are all taken from the same population. |
| 6.7i | A | Parthenogenesis is the production of offspring by an animal without the fertilisation of an egg. |
| 6.8i | B | Members of the Camelidae (camels, llamas and their relatives) do exhibit induced ovulation. |
| 6.9i | D | Signs of imminent parturition vary between species but all of the listed signs are common. |
| 6.10i | C | This is the definition of mean kinship. It is assessed in order to preserve genetic diversity in small populations. |
| 6.11i | D | The coefficient of relatedness ranges from zero (unrelated) to 1.0 (identical). Clones are genetically identical. |
| 6.12i | A | This population is decreasing because it contains very few young animals. |

| 6.13i | A | The effective population size ($N_e$) is the same as the actual population size when the sex ratio is 1:1, assuming all of the individuals are capable of reproduction. In a population of 100 (50 male and 50 females) the effective population is also 100:<br><br>$$N_e = \frac{4(50 \times 50)}{(50 + 50)} = 100$$ |
|---|---|---|
| 6.14i | B | Female birds possess two different types of sex chromosome, hence they are heterogametic. |
| 6.15i | C | Elephants exhibit two surges in LH – rather than the usual single surge – and it is the second surge that induces ovulation. |
| 6.16i | B | Most female birds have one functional oviduct not two. |
| 6.17i | A | A PCR machine uses the polymerase chain reaction (PCR) to amplify small segments of DNA to produce a sufficient quantity for analysis. |
| 6.18i | D | SPARKS is an acronym for Single Population Analysis and Record Keeping System. |
| 6.19i | C | Genetic drift is the variation in the relative frequency of different genotypes in a small population resulting from the random loss of particular alleles due to chance events such as natural disasters. |
| 6.20i | A | A is correct. The distracters are the same terms matched with incorrect definitions. |
| 6.1a | D | D is correct because all of the other options are true: docile animals are easier to manage, have adapted well to captivity and are less likely to be harmed or harm themselves. |
| 6.2a | B | Inbreeding results in an accumulation of deleterious genes. This is called genetic load. Mutation rate is not affected by inbreeding. |
| 6.3a | D | TRPV4 is the protein that results in the phenomenon whereby environmental temperature determines the sex of the offspring produced. |
| 6.4a | C | Minimum viable population is the smallest number of individuals required for a population or species to have a specified probability of surviving for a given period of time. Zoos often use a minimum effective size of 50 breeding individuals as the minimum short-term size to keep inbreeding below 1% per generation. A size of 500 is considered necessary to reduce genetic drift. This so called 50/500 rule may not be widely applicable across taxa. |
| 6.5a | D | Individuals C and E are unrelated so $r = 0$. |
| 6.6a | C | *Penelope* has half of *Anita's* alleles and *Lindo* has half of *Penelope's* alleles, so the *r* value is 0.5 x 0.5 = 0.25. They are grandmother and granddaughter. |

| 6.7a | D | Each of these relationships produces a value of $r = 0.25$ |
|------|---|---|
| 6.8a | D | All of the first three options will increase effective population size because these management practices will reduce inbreeding and help to maintain a 1:1 sex ratio thereby keeping the number of breeding animals at a maximum. |
| 6.9a | D | Between the ages of 2 years and 3 years 137 out of 987 died so the age-specific death rate is 137/987 (i.e. 0.139). |
| 6.10a | B | In monomorphic species the sexes look the same. A laparoscope is an optical instrument that may be used to examine the internal organs of a bird to determine, among other things, its sex. |
| 6.11a | B | Prolactin is a hormone released from the anterior pituitary gland at the base of the brain and stimulates milk production. |
| 6.12a | C | In harem species, such as many deer and antelopes, one male takes control of a group of females thereby excluding other males from breeding. This creates a problem for zoos because of the number of excess male animals born into the populations of harem species. |
| 6.13a | A | A copulatory tie is a phenomenon whereby the male is temporarily unable to separate from the female after mating and is characteristic of canids. |
| 6.14a | B | A feather taken from the cage floor could be contaminated with DNA from another bird and may not contain any cells suitable for extracting DNA. |
| 6.15a | C | Males = 12+6=18; females = 29+7 = 36. The ratio 18:36 can be simplified to 1:2. |
| 6.16a | A | The studbook keeper manages breeding for a particular species held in zoos that contribute to a breeding programme. The use of inbreeding coefficients helps to determine which individuals in the population should not be used for further breeding to minimise inbreeding. |
| 6.17a | D | The format is males.females.unknown sex. |
| 6.18a | D | The number of males = 7-0+1+1-0 = 9; females = 4-1+2+2-1 =6; unknown sex = 2-0+0+0-1 = 1; total at the end of the year = 9.6.1. |
| 6.19a | C | If the recessive allele for albinism is $a$ an albino female would have the genotype $aa$. A normal male whose mother was an albino would have the genotype $Aa$. If an $Aa$ individual mates with an $aa$ individual the probability that an $aa$ individual will be produced from this cross is 0.5. |
| 6.20a | B | The alleles $r$ and $s$ are sex-linked because they are located on the non-homologous section of the X chromosome. |

# Chapter 7 Conservation Breeding and Assisted Reproductive Technologies

| 7.1f | B | Cryobiology is correct. Cryptozoology is the study of lost and unknown animals e.g. the yeti, the Loch Ness monster. |
|------|---|---|
| 7.2f | C | *In-vitro* means, 'within the glass', in Latin. *In-vitro* fertilisation occurs in the laboratory. |
| 7.3f | D | The fish hawk or osprey is a common and widely distributed bird of prey and classified by the IUCN as 'Least Concern' on its Red List. All of the other species have been saved from extinction by captive breeding. The Mauritius kestrel and Przewalski's horse are classified as 'Endangered'; the California condor is classified as 'Critically Endangered'. |
| 7.4f | B | In the wild the female nests in a tree hole. This is then partially sealed with mud and she is fed by her partner through the hole. |
| 7.5f | C | *Patula* is a genus of tropical land snails. |
| 7.6f | A | A frog tadpole (*Xenopus*) was cloned by John Gurdon in 1958 at the University of Oxford. |
| 7.7f | A | EEPs are the most intensive population management programmes (breeding programmes) operated by zoos belonging to the European Association of Zoos and Aquaria (EAZA). The acronym EEP originally meant European Endangered Species Programme but now applies to EAZA *Ex-situ* Programmes. |
| 7.8f | B | The WWT is the Wildfowl and Wetlands Trust and, among other things, breeds rare waterbirds (especially ducks and geese). |
| 7.9f | D | An ESU is an evolutionarily significant unit: a group of organisms considered distinct for conservation purposes. |
| 7.10f | B | A metapopulation is a group of geographically separated organisms of the same species that interact, especially by interbreeding. |
| 7.11f | C | The widespread destruction of the bison (buffalo) in North America was one of the reasons why the National Zoological Park was created. |
| 7.12f | C | The logo used by the SSP programmes in North America consists of an adult rhinoceros with a calf. |
| 7.13f | D | Inbreeding depression is the reduction of genetic fitness that occurs when close relatives mate and there is an accumulation of deleterious recessive alleles in the gene pool. |
| 7.14f | B | This is double clutching and increases fecundity (reproductive output). |

| 7.15f | D | The AZA captive breeding programmes are called Species Survival Plans. The distracters are fictitious in this context. |
|---|---|---|
| 7.16f | A | Tubal ligation is a contraceptive surgical procedure for females whereby the Fallopian tubes are cut or blocked to prevent pregnancy. |
| 7.17f | D | The Chinese Government loans giant pandas to suitable zoos around the world for payment on the condition that any cubs produced remain the property of China. In the past giant pandas were sometimes given by China to foreign governments as a gesture of friendship and goodwill. |
| 7.18f | B | This process is called cloning. Cleavage is the division of cells during embryonic development. |
| 7.19f | D | All of the listed species have been bred in captivity and then released back into the wild. |
| 7.20f | A | Studbooks exist for all of these species. The first was for the European bison. |
| 7.1i | B | If individuals of a particular species breed at the age of 10 years in the wild but are bred at the age of 5 years in zoos the generation time is halved. There is evidence that some genetic diversity is lost with each generation in small populations. Breeding at age 5 years will increase the rate of growth of the population but will lose genetic diversity faster than if the animals were bred at age 10 years. |
| 7.2i | D | The quagga was a type of zebra previously found in southern Africa and hunted to extinction in the 19th century: *Equus quagga quagga*. |
| 7.3i | B | A leopon is a cross between a male leopard and a female lion. As an interspecies hybrid it has no conservation value. The distracters are rare species and subspecies. |
| 7.4i | B | Most of the animals kept in most of the zoos in the world will die in a zoo. A relatively small number of species are currently being reintroduced into the wild and sometimes such reintroductions meet with limited success. This was particularly the case with early attempts at reintroduction. |
| 7.5i | C | Headstarting increases the chances that released individuals will survive. |
| 7.6i | A | The One Plan Approach is an attempt to integrate *in-situ* and *ex-situ* conservation efforts for their mutual benefit. The distracters are fictitious in this context. |
| 7.7i | C | This baby gorilla was born to a western lowland gorilla, Rosie, at Cincinnati Zoo in 1995. The sperm used was obtained from a male at Omaha's Henry Doorly Zoo. |

| 7.8i | D | Cloning is expensive, the individuals produced are, by definition, genetically identical to other individuals and health problems are common in neonates. |
|---|---|---|
| 7.9i | C | The candle test (candling) consists of holding a bright light behind an object to create a silhouette of its contents e.g. a male seahorse's egg pouch or a bird's egg. |
| 7.10i | B | The World Association of Zoos and Aquariums (WAZA) maintains the international studbooks for rare and endangered species. |
| 7.11i | D | The tarsus on their most caudal legs is used to prop up the female. For copulation they also need complete pedipalps and their first pair of legs. |
| 7.12i | A | The Hawaiian goose was captive-bred and released in Hawaii as a result of the efforts of the Wildfowl Trust (now the Wildfowl and Wetlands Trust) working with the Honolulu Zoo and others. |
| 7.13i | B | There are too few Asian elephants in captivity in zoos to produce viable breeding programmes for the separate subspecies. Furthermore, the captive elephant population includes some hybrids that would need to be excluded from further breeding if the captive population were to be subdivided into subspecies. |
| 7.14i | A | A is correct. The distracters are fictitious in this context. |
| 7.15i | C | *PopLink* is correct. The distracters are fictitious in this context. |
| 7.16i | B | This phenomenon is hybrid vigour and results in the production of more heterozygotes. Inbred populations tend to contain individuals that are homozygous for many genes making it more likely that abnormalities and disease will be manifested. |
| 7.17i | D | A flow cytometer can be used to sort sperm based on the sex chromosomes they carry. |
| 7.18i | B | Vitrification is the rapid cooling of a liquid without the formation of ice crystals that could damage cell structure. |
| 7.19i | B | Lifetime reproductive planning for a species must balance the benefits of early and regular reproduction (which produces more individuals) with the loss of genetic diversity in each generation, changes to the population age structure (which may adversely affect social structures) and the capacity of the zoo community to accommodate more animals. |
| 7.20i | A | MGA is a contraceptive. |
| 7.1a | B | B is correct. The distracters are the same organisations matched with the incorrect gene banks. |
| 7.2a | A | Sperm is stored using liquid nitrogen to keep it at -196°C. |

| 7.3a | C | Female mammals may be induced to mature and release more eggs than usual by treatment with sex hormones (e.g. luteinising hormone). In the captive breeding of rare mammals these eggs may subsequently be used in IVF and embryo transfer or frozen and stored for future use. |
|------|---|------|
| 7.4a | D | The Association of Zoos and Aquariums has a Wildlife Contraception Center at St Louis Zoo in Missouri. |
| 7.5a | B | GnRH is gonadotropin-releasing hormone. Slow-release implants of GnRH are used to suppress oestrus and prevent reproduction. |
| 7.6a | A | The world's first interspecies frozen/thawed embryo transfer resulted in the cloning of an African wild cat (*Felis silvestris lybica*) at the Audubon Nature Institute, Louisiana, United States, in 1999. |
| 7.7a | A | Evolutionarily significant unit (ESU) is the correct answer. It has been suggested that these two populations should be treated as different species. |
| 7.8a | D | Of the organisations listed, only the Audubon Nature Institute has not been involved in the recovery plan for black-footed ferrets. |
| 7.9a | C | The cockatiel was being used as a prototype species. |
| 7.10a | B | Microsatellite markers is correct. The distracters are fictitious in this context. |
| 7.11a | C | Mountain gorilla populations have recovered as a result of the use of interventions in the wild (*in-situ*). There is no *ex-situ* captive breeding programme for these animals. |
| 7.12a | A | The gaur is a wild ox from Southeast Asia. The calf survived for just two days. |
| 7.13a | A | This process is called cross-fostering: surrogate parents from a related common species are used to rear the young of a rare species. |
| 7.14a | D | The domestic horse is used as a surrogate species in this process and will give birth to a Przewalski's horse foal. |
| 7.15a | C | C is the correct sequence. The distracters are the same elements in incorrect sequences. |
| 7.16a | B | Although there may be instances where the ZSL is involved in wildlife reintroductions the society has no formal role in reintroductions in general. |
| 7.17a | A | CREW is located at Cincinnati Zoo in Ohio, United States. |
| 7.18a | D | Zoos would not normally use contraception to assist in the selective breeding for certain characteristics in a population (except perhaps to prevent the inheritance of a congenital condition or disease). Selecting for tameness could jeopardise future reintroduction plans by producing individuals with no fear of humans. |

| 7.19a | C | It is widely accepted within the zoo community that species should be managed so as to preserve 95% of genetic diversity in a demographically stable population for 200 years. |
|---|---|---|
| 7.20a | A | SCNT is somatic cell nuclear transfer. This is a method of producing a viable embryo in the laboratory from an egg cell and the nucleus from a body (somatic) cell. |

# Chapter 8   Behaviour, Training and Environmental Enrichment

| 8.1f | C | An ethogram is a list and description of all the behaviours exhibited by a particular species or a subset of those behaviours of interest in a particular study. It may include photographs or diagrams of behaviours. |
|---|---|---|
| 8.2f | B | Shepherdson's definition states that enrichment should contribute to the psychological and physiological wellbeing of animals. |
| 8.3f | A | When animals are tame they do not take flight when approached by humans. |
| 8.4f | C | Individuals of the same species living in geographically separated areas in the wild develop different traditions including those related to the selection and handling of food. This occurs, for example, in relation to tool use in chimpanzees (*Pan troglodytes*). |
| 8.5f | A | Chimpanzees (*Pan troglodytes*) make these sounds and many others. |
| 8.6f | B | Dr Hal Markowitz pioneered behavioural engineering and is the author of *Behavioral Enrichment in the Zoo*. |
| 8.7f | D | D is correct. The distracters are the same terms matched with the incorrect times of day. |
| 8.8f | D | Enrichment has the potential to promote natural behaviour, allow an animal control over its environment and hence enhance wellbeing. |
| 8.9f | C | The sense of taste is gustation. |
| 8.10f | B | Zoos sometimes provide enrichment for animals such as primates and elephants by giving them fruit and fruit juice in a block of ice. |
| 8.11f | C | A *Boomer Ball* is a proprietary make of large ball given to some animals as enrichment. Note the scratches on this ball which was located in a tiger enclosure. |
| 8.12f | B | Giraffes are prone to developing an oral stereotypy whereby they repeatedly lick metal bars in their enclosures. |

| 8.13f | C | Vacuum behaviour is a stereotyped sequence of activities that occurs in the absence of the appropriate stimuli, e.g. birds trying to 'catch' imaginary insects and raccoons 'washing' food in the absence of water. |
|-------|---|---|
| 8.14f | A | SIB stands for self-injurious behaviour and includes excessive grooming (sometimes to the point of baldness), self-biting and feather plucking. |
| 8.15f | D | Goad, ankus and bullhook are all terms for this tool. The use of an ankus in the handling of elephants has been banned in some jurisdictions. |
| 8.16f | B | Shaping is the process whereby behaviour is gradually modified by providing rewards each time the animal gets a step closer to performing the desired act. For example if the desired act is for a rhinoceros to touch a target with his nose the rhino should be rewarded with a small piece of food each time he takes a step closer to the target but not when he takes a step away from it. |
| 8.17f | A | A lek is a place where the males of certain bird species gather to attract and mate with females. Leks are also used by some antelope species. |
| 8.18f | D | Habituation occurs when an animal is repeatedly exposed to a potentially aversive stimulus and learns that it is not harmful so stops responding to it. |
| 8.19f | A | An ethogram may consists of a subset of the species' behavioural repertoire. For example, if a scientist is studying social behaviour the ethogram may only contain definitions of social behaviours. |
| 8.20f | C | Pheromones are known to mediate sex and dominance in aquatic organisms including fishes, crustaceans and polychaetes. |
| 8.1i | D | These are precocial chicks. Altricial chicks hatch in a helpless state. |
| 8.2i | B | This is behavioural restriction and occurs when animals are kept in poor conditions and in appropriate social groups. The distracters are fictitious in this context. |
| 8.3i | B | This test involves surreptitiously placing a red spot (or similar mark) on the face of an animal and then watching its response to this when looking in a mirror. If the animal touches the coloured spot more often than anywhere else on its face it is assumed that it recognises that the image in the mirror is its own face and not that of another individual. A number of studies of this phenomenon have been conducted using animals in zoos. |
| 8.4i | A | Operant conditioning is also called instrumental conditioning or instrumental learning. Classical conditioning involves an association between an involuntary response and a stimulus. |

| 8.5i | B | Saltatory locomotion (saltation) involves moving by leaps and jumps. |
|---|---|---|
| 8.6i | C | This method of elephant handling does not use an ankus or any coercive techniques. |
| 8.7i | A | This is a definition of animal personality. |
| 8.8i | D | This a stereotypy (stereotypic behaviour). The distracters are other animal behaviour terms. |
| 8.9i | C | This is a quantitative change as it is a change in the frequency of an existing behaviour. |
| 8.10i | D | Instrumental learning – also known as operant conditioning – is used to train animals to move between different elements of their enclosures, cooperate with veterinary examinations and enter crates for transportation. |
| 8.11i | C | These straps are known as jesses and are used to tether a bird to a post or stand when not being flown or control it while in the hand of a falconer or keeper. |
| 8.12i | D | This is cognitive enrichment because it is mentally challenging. The distracters are fictitious in this context. |
| 8.13i | B | A 'wobble tree' is an enrichment device consisting of a dead or artificial tree (or a platform fixed to the top of a long pole) that moves when shaken and releases food. A red-legged seriema is a bird. |
| 8.14i | A | A piñata is a container filled with food treats that may be given to animals living in zoos as a form of enrichment. To gain access to the food the container must be broken or torn open. Brightly-coloured piñatas are associated with Mexico. |
| 8.15i | C | This is a sociogram where each dyad (pair of animals) is linked by a line whose thickness indicates the strength of association. Thick lines join close associates. |
| 8.16i | B | *Upali* – a juvenile bull – is the least social because, although he associates with all of the other elephants, each of the lines is thin. |
| 8.17i | D | When an animal is being trained it may not be possible to reward it at the precise moment it performs the desired behaviour. The sound of a clicker is used to mark the time when the desired behaviour occurs and this is linked to the reward that follows. |
| 8.18i | C | This is operant conditioning or instrumental learning. |
| 8.19i | A | The food reward functions as positive reinforcement. The distracters are fictitious in this context. |
| 8.20i | B | The proximate cause is the event that is immediately responsible for something (such as a particular behaviour) that has happened. |

| 8.1a | D | D is correct. The distracters are fictitious in this context. |
|------|---|---|
| 8.2a | C | If the time when the animal was out of sight is excluded (3 recordings) the total number of recordings was 36-3 = 33. The percentage of time spent feeding was (18/33)/100 = 54.5%. |
| 8.3a | B | Head-butting is a natural behaviour for bison and a head-butting post provides them with an opportunity to do this. A burlap bag is a sack made from hessian or a similar fabric that may be filled with hay or other foods. |
| 8.4a | D | All of the options may result in differences in the way that animals use their enclosures. |
| 8.5a | C | Target training is a type of operant – not classical – conditioning. |
| 8.6a | B | Perseveration is the performance of an activity when the appropriate stimulus is absent or when it has ceased. |
| 8.7a | A | Bill Travers was the co-founder of Born Free and he used this term to describe the repetitive behaviour observed in animals in zoos. The distracters are well-known experts in animal welfare. |
| 8.8a | D | Neophobia is a fear of new objects or experiences. Neophilia is an attraction to novel things. Xenophobia is a fear or dislike of strangers or foreigners and is a human trait. Necrophilia is a sexual attraction to dead bodies. |
| 8.9a | B | Many captive animals appear to prefer to work for their food even when it is freely available. This behaviour is called contrafreeloading. The term 'freeloading' means taking advantage of the generosity of others and giving nothing in return. |
| 8.10a | A | SPIDER is an acronym relating to a model for establishing environmental enrichment programmes: Setting goals, Planning, Implementation, Documenting, Evaluation, Readjustment. |
| 8.11a | B | B supplanted all of the other animals (A 24 times, C 17 times and D 9 times). None of the other animals supplanted B. |
| 8.12a | D | In relation to behaviour, the term 'ontogeny' relates to the development of a particular behaviour from the earliest stage to maturity. |
| 8.13a | C | Maintaining behavioural competence in captive-bred animals reared for conservation purposes is essential, especially if they are to be released to the wild. |
| 8.14a | A | The male stands behind the female and uses one of his forelegs to tap one of her hind legs during courtship. |
| 8.15a | B | When an animal stops responding to a harmless stimulus it has undergone habituation. When an animal becomes especially sensitive to a stimulus sensitisation has occurred. |

| 8.16a | A | Learned behaviours become extinct if not reinforced. |
| 8.17a | D | Pursuing food organisms is a type of appetitive behaviour; consummatory behaviour satisfies a specific drive, in this case eating captured prey. |
| 8.18a | A | In the wild animals spend a greater deal of time on activities necessary for survival such as searching for food. Captive animals have more time for leisure activities such as play. |
| 8.19a | C | Brumation is a type of dormancy or torpor in response to cold temperatures. In very hot weather aestivation may occur. |
| 8.20a | A | Courtship has all of these functions. |

# Chapter 9 Animal Welfare and Conservation Medicine

| 9.1f | A | Preventative medicine (for example regular veterinary examinations, routine foot care, and vaccination) is important in detecting disease at an early stage and preventing contagious disease from spreading. |
| 9.2f | C | This argument claims that it is acceptable to keep an animal in a zoo only if its life there is at least as good as it would be in the wild. |
| 9.3f | A | Animals should be introduced to the social group gradually to insure that they are compatible with the existing members. Individuals are normally held in quarantine before being introduced to an established group to prevent the introduction of contagious diseases and parasites. However, there is a move to be more flexible with quarantine rules, especially when animals are moved between institutions with high animal health standards and between two zoos belonging to the same overarching institution. |
| 9.4f | A | These are signs. Animals do not have symptoms because a symptom is subjective and only apparent to the patient. An animal cannot describe to us how it feels. A sign is something that can be objectively appreciated, e.g. a broken bone, an abnormally high level of calcium in the blood, a lump or skin condition. |

| 9.5f | B | Euthanasia of an animal may be justifiable for a number of reasons. An animal may be very old and infirm, ill, blind, unable to feed or have some other chronic incurable condition. It may be in great pain. Sometimes animals may be surplus to a zoo's requirements and it may be impossible to place them elsewhere. A zoo may close (or be closed) and it may not be possible to rehome the animals. An animal that is incapable of breeding or whose genes are over-represented in the population may be euthanised because it is using accommodation that could be used for a breeding animal. |
|---|---|---|
| 9.6f | A | A shift box is a small container used to transport snakes, especially venomous snakes, safely and securely. |
| 9.7f | D | A malocclusion is a misalignment of the teeth (dental malocclusion) or the whole jaw (skeletal malocclusion). |
| 9.8f | C | A transponder is inserted under the skin which contains information about the identity of the individual. |
| 9.9f | B | Stress causes a build up of lactic acid. This causes a breakdown of cell membranes releasing the cell contents. Animals may suffer from shock, electrolyte imbalance, or muscle damage and may die as a result. |
| 9.10f | D | Analgesics are painkillers. |
| 9.11f | C | Morbidity is a measure of the amount of disease in a population. It is not the same as mortality which relates to deaths. |
| 9.12f | B | This is a chronic disease because the ape has lived with it for a long time and requires ongoing veterinary attention and dietary adjustments. Acute disease has a rapid onset and requires urgent or short-term treatment. |
| 9.13f | D | This bear is following (tracing) the same route around its enclosure repeatedly; this is repetitive and has no clear purpose so is a stereotypic behaviour; it also an abnormal behaviour. |
| 9.14f | B | Pressure platforms are used for gait analysis and measure the pressure under different parts of the foot. |
| 9.15f | C | The genes of this giraffe were over-represented in the captive population at the time so he could not be used for breeding. |
| 9.16f | A | There is some evidence that the presence of stereotypic behaviour in animals is accompanied by changes to the physical structure of the brain. |
| 9.17f | D | Longevity and life span mean the same thing: the length of life from beginning to death. Life expectancy is the average number of years an individual of a particular age is likely to live. It is a prediction of the number of years left in an individual's life. |

| 9.18f | B | UVB light is a type of ultra-violet light with a wavelength of 280 −315nm. It is important in the synthesis of vitamin $D_3$ which assists in the absorption of calcium. |
|---|---|---|
| 9.19f | C | If an animal can be approached to within a short distance it may be possible to inject it with a syringe attached to the end of a pole. A blowpipe may be used over a slightly longer distance. At greater distances a dart pistol or rifle is required. |
| 9.20f | B | Hypercalcaemia in lizards may cause bone defects, cardiac abnormalities, shock and sometimes renal failure and death. Hypocalcaemia is a lack of calcium; hypoglycaemia is low blood sugar level; hyperhidrosis is excessive sweating and could not occur in a reptile. |
| 9.1i | D | Pododermatitis is inflammation of the skin between the toes and footpads. Although often associated with penguins and flamingos in zoos, this disease can also affect rodents, rabbits and other species, including mink and dogs. |
| 9.2i | C | Dystocia is a difficult or obstructed labour resulting, for example, from an abnormally large foetus or the foetus being incorrectly positioned. |
| 9.3i | D | Cleaning the teeth and chewing materials such as bark can help to prevent and treat periodontal (gum) disease. |
| 9.4i | B | Psittacosis is predominantly a bacterial disease of birds. However it has been reported from some mammals and affects humans. |
| 9.5i | C | Helminths are platyhelminths or flatworms such as flukes or tapeworms. |
| 9.6i | D | West Nile virus is a mosquito-borne disease that primarily affects birds but can also affect a range of other species including humans. |
| 9.7i | B | Spondylosis is a type of arthritis that affects the spine. It is caused by wear and tear and can result in restricted movement of the vertebrae. |
| 9.8i | C | Rabies only affects mammals. |
| 9.9i | A | A nebuliser is a device for delivering a drug that needs to be inhaled by converting a solution into a fine spray. |
| 9.10i | D | Stress could have any of these effects. |
| 9.11i | B | Autophagy means 'self-eating'. Autotomy and autotilly refer to voluntary self-amputation of an appendage (e.g. a lizard losing its tail to escape from a predator). Autolysis is the process of self-digestion of cells by their own digestive enzymes. |

| 9.12i | A | Emesis is the action of vomiting and may occur as a result of motion sickness. |
|---|---|---|
| 9.13i | B | Laminitis is a condition that affects hooves and is found mostly in equids and bovids. |
| 9.14i | C | Revivon (diprenorphine) is used as an opioid antagonist to reverse the effect of opioid analgesics such as etorphine. The distracters are all drugs used to immobilise animals. |
| 9.15i | C | There are many different body condition scoring systems for animals. Scores typically range from 1-5 or 1-9. |
| 9.16i | D | The disease can be transmitted to humans but is rare. In humans it is not the same as the hand, foot and mouth disease common in children. |
| 9.17i | B | Cortisol is often known as the 'stress hormone'. |
| 9.18i | A | A crush (or squeeze) is a device for restraining a large animal, usually for a veterinary procedure or examination. |
| 9.19i | A | Feline panleukopenia (feline distemper) affects felids and related taxa of mammals such as dogs, foxes, skunks and raccoons. |
| 9.20i | C | Many birds naturally bob their tails. However, excessive tail-bobbing is a sign of respiratory disease. |
| 9.1a | B | This is lumpy jaw or Actinomycosis and is caused by the bacterium *Actinomyces bovis*. |
| 9.2a | D | In the United Kingdom, the United States, Canada, New Zealand, Australia and some other jurisdictions such diseases are called notifiable diseases. |
| 9.3a | C | Coprophagia, or coprophagy, is the eating of faeces. It occurs naturally in some species, e.g. rabbits, but is abnormal in most species. |
| 9.4a | B | HVAC is an acronym for Heating, Ventilation and Air Conditioning. This is especially important if animals are likely to be exposed to extremes of temperature during transportation or are being moved between different climatic zones. |
| 9.5a | C | This is a controversial report produced in 2002. It was subtitled *How elephants suffer in zoos* and called for the phasing out of elephants in European zoos. |
| 9.6a | A | A luxation is a joint dislocation. In this case it is the hip joint. |
| 9.7a | D | Locations 2 and 3 are areas of thick muscle suitable for receiving a hypodermic dart. |
| 9.8a | C | A tiger at the Bronx Zoo was the first animal living in a zoo in the United States to be diagnosed with COVID-19. |

| 9.9a | A | Collapsed dorsal fins are frequently seen in captive orcas (killer whales). The cause is unclear and it is particularly prevalent in males. |
|------|---|---|
| 9.10a | C | *Ichthyophthirius multifiliis* is a ciliate that causes small white spots over the body. |
| 9.11a | D | Usable volume is correct. For many species the height (or depth in aquatic species) provides important usable space. |
| 9.12a | C | Chytridiomycosis is a fungal disease that affects amphibians. |
| 9.13a | A | An elevator is a tool used in dental extractions to loosen teeth. |
| 9.14a | C | This involves floating eggs to the surface of a suspension of faeces and counting the eggs under a grid on a McMaster slide using a microscope. |
| 9.15a | C | C is false. Some individuals within a species are better able to cope with stress than others. Different strains of inbred mice have been shown to exhibit dramatically different behavioural and physiological responses to stress. |
| 9.16a | B | It is not possible to say whether or not the animals experienced more stress on cold days than on warm days as the level of stress they experienced was not measured. There appears to be an association between temperature and the frequency of stereotypic behaviour. To investigate this association it would be necessary to calculate a correlation coefficient between the temperature on specific days and the frequency of stereotypic behaviour on those days. Even if this found a high negative correlation between temperature and the frequency of stereotypic behaviour it cannot be inferred that this frequency was a direct result of temperature. In any event, it would be necessary to show that the frequency of stereotypic behaviour could be used as a proxy for measuring stress in this species. |
| 9.17a | D | Cortisol can be measured from samples of blood, saliva, urine or faeces. |
| 9.18a | C | Being kept alone, fed at predictable times and being able to see other cheetahs in adjacent enclosures all increased the prevalence of stereotypic behaviours. |
| 9.19a | B | An ocular coloboma is a congenital eye malformation: a gap which may be in the iris, lens, retina or elsewhere in the eye. |
| 9.20a | C | Some zoos use a risk-based approach to minimise or eliminate quarantine for animals transferred from institutions that operate a comprehensive disease surveillance system. |

# Chapter 10   Zoo Organisation and Regulation

| 10.1f | A | The Zoological Society of London has buildings adjacent to London Zoo in Regent's Park. |
|---|---|---|
| 10.2f | B | CITES as an acronym for the Convention on International Trade in Endangered Species of Wild Fauna and Flora 1973. |
| 10.3f | D | Species360 is correct. The distracters are fictitious. |
| 10.4f | C | This is the Association of British and Irish Wild Animal Keepers. |
| 10.5f | B | An endocrinologist studies hormones. The nutritional value of food would be the concern of a nutritionist. |
| 10.6f | D | The licensing of zoos varies between countries. In the United Kingdom it is the responsibility of Local Authorities (local government), in Australia it is the state government and in India it is the central government. |
| 10.7f | B | The institute is internationally important in zoo research and is located in Berlin, Germany. |
| 10.8f | A | A is correct. The distracters are fictitious. |
| 10.9f | C | The IUCN maintains a Red List indicating the conservation status of a large number of species and updates this periodically. |
| 10.10f | C | C is correct because the rhinoceroses are being conserved in a place where they naturally occur. All of the distracters are *ex-situ* projects because the animals are being bred in a zoo or aquarium. |
| 10.11f | A | The International Species Inventory System was renamed the International Species Information System and later formed the basis of information held by Species360. |
| 10.12f | B | AZA accredited its first zoo in 1974. |
| 10.13f | C | A computational biologist would be best qualified for this type of work. Each of the distracters embraces a much wider range of expertise. |
| 10.14f | B | Traditionally a zoo would have a curator of mammals, a curator of birds and curators of other particular taxa. The same title is used for someone responsible for a particular collection of items in a museum. |
| 10.15f | D | CITES covers living and dead animals and recognisable parts of animals including objects containing, or made from, animal parts. |
| 10.16f | D | Most of the institutions for which the Association of Zoos and Aquariums (AZA) provides accreditation are located in North America. However, it also accredits facilities in Mexico, Singapore, South Korea, Argentina and a number of other countries. |

| 10.17f | C | IATA is the International Air Transport Association and sets global standards for animal transportation in its Live Animal Regulations. |
|---|---|---|
| 10.18f | A | Zoos in the EU must comply with the Zoos Directive. This requires zoos to keeps records of their animals. |
| 10.19f | A | The European Union of Aquarium Curators was created in 1972 to ensure closer contact between curators in the future. |
| 10.20f | C | In 2020 the AZA provided accreditation for 240 zoos and aquariums. |
| 10.1i | D | ZIMS is the Zoological Information Management System. The distracters are fictitious. |
| 10.2i | A | In the European Union a zoo open to the public for 7 or more days in any year requires a licence under the Zoos Directive unless it is exempt. |
| 10.3i | B | A collection plan shows the species a zoo keeps and the species it intends to keep in the future. |
| 10.4i | D | The person in a zoo that keeps the animal records is generally called the registrar. |
| 10.5i | A | The ZooLex Zoo Design Organization is a non-profit organisation established to improve holding conditions for wild animals kept in captivity by improving zoo design. |
| 10.6i | B | The Conservation Grants Fund was created by the Association of Zoos and Aquariums in 1984. |
| 10.7i | C | The Animal and Plant Health Inspection Service (APHIS) is part of the US Department of Agriculture (USDA) and is responsible for zoo inspections in the United States. |
| 10.8i | C | A figure of 700 million visitors appeared on WAZA's website (in 2021) but this same figure also appeared in previous years. It refers to visits rather than different individuals so is somewhat misleading. |
| 10.9i | D | The Wildlife Conservation Society is the parent body of the Bronx Zoo in New York. |
| 10.10i | C | In the European Union, the Balai Directive is concerned with the export, import, and movement of live animals and germplasm (sperm and ova). |
| 10.11i | B | *Committing to Conservation* was published in 2015 and is the conservation strategy of the World Association of Zoos and Aquariums (WAZA). |
| 10.12i | A | Herpetology is the study of amphibians and reptiles. |

| 10.13i | D | In England zoo inspectors are appointed by the Secretary of State (currently the Secretary of State for the Department for Environment, Food and Rural Affairs (Defra)) and, for any particular zoo, by the Local Authority of the area in which the zoo is located. |
|---|---|---|
| 10.14i | B | In 2020, there were 310 licensed zoos and aquariums in the United Kingdom according to BIAZA. |
| 10.15i | A | The projects listed were initiatives of the European Association of Zoos and Aquaria. |
| 10.16i | C | Under the authority of the Zoo Licensing Act 1981 the Secretary of State produces guidance for zoos as the *Secretary of State's Standards of Modern Zoo Practice*. |
| 10.17i | D | The Central Zoo Authority is the agency responsible for zoo licensing in India. PETA is People for the Ethical Treatment of Animals and CAPS is the Captive Animals' Protection Society. These organisations and the Born Free Foundation campaign against the existence of zoos. |
| 10.18i | D | D is the correct order. The distracters contain the same terms in incorrect orders. |
| 10.19i | A | The Red List Index is based on the IUCN's Red List of Threatened Species. Zero indicates all species of a particular taxon have been lost; 1.0 equates to all species categorised as 'Least Concern' and therefore none are expected to become extinct in the near future. |
| 10.20i | B | The trend for many large zoos to construct large, expensive, themed exhibits has been called 'Disneyization'. |
| 10.1a | B | CITES lists protected species in Appendices I, II and III. |
| 10.2a | C | The Species Conservation Toolkit Initiative (SCTI) is an international initiative that involves a large number of organisations including those in the distracters. |
| 10.3a | D | Taxon Advisory Groups vary considerably in the taxa they cover. TAGs managed by the AZA include the Amphibian TAG (for a class), the Felid TAG (a family) and the Bat TAG (an order). They also include other groupings of taxa such as New World primates, small carnivores, freshwater fishes and terrestrial invertebrates. |
| 10.4a | D | The head office of EAZA is located in Artis Zoo, Amsterdam. |
| 10.5a | B | The International Association of Directors of Zoological Gardens first met in 1935. It changed its name to the International Union of Directors of Zoological Gardens in 1946. In 1992 this became the World Zoo Organisation and in 2000 this became the World Association of Zoos and Aquariums (WAZA). |

| 10.6a | C | ARKS is short for Animal Record Keeping System and is part of a larger system called Species360 ZIMS. |
|---|---|---|
| 10.7a | B | Zoos and aquariums must comply with the national legislation relating specifically to zoos in the country (and/or state) where they are located. In addition they must comply with more general legislation concerning animal welfare, health and safety, the disposal of hazardous waste, the prevention of the spread of disease, employment, etc. |
| 10.8a | A | A is the correct sequence, beginning in 1995 and ending in 2021. Note that as time progressed the mission became increasingly more complex and then (after 2004) the message was simplified to just two words: '…preventing extinction'. |
| 10.9a | D | The Aichi Targets were established under the UN Convention on Biological Diversity. |
| 10.10a | A | The international Species Inventory System was established by Dr Ulysses Seal in 1973. |
| 10.11a | C | A Regional Collection Plan is made by a Taxon Advisory Group for a specific taxon of animals or a group of taxa. |
| 10.12a | B | Article 9 of the United Nations Convention on Biological Diversity 1992 requires that *ex-situ* conservation measures should be adopted for components of biodiversity, preferably in their country of origin. |
| 10.13a | D | The Central Zoo Authority oversees the operation and licensing of zoos in India. |
| 10.14a | B | In 1974, 55 zoos in the United States and Europe joined ISIS as the result of a proposal by its founder, Ulysses Seal. |
| 10.15a | A | *MedARKS* is an abbreviation for Medical Animal Record Keeping System. |
| 10.16a | D | ZIMS for *Ex-situ* Conservation does not exist. |
| 10.17a | A | The CITES Management Authority of a Contracting Party to CITES issues this documentation. |
| 10.18a | C | LaCONES is located in Hyderabad, India. |
| 10.19a | C | ZIMS was first rolled out as a web-based system in 2011. |
| 10.20a | D | By 2018 ZIMS was operating in English, Spanish, Russian and Japanese. |

# References

Beardsworth, A. and Bryman, A.E. (2001) The wild animal in late modernity. The case of the Disneyization of zoos. *Tourist Studies* 1, 83-104.

Bitgood, S. (1988) Problems in visitor orientation and circulation. In: Bitgood, S., Roper, J. and Benefield, A. (eds) *Visitor Studies – 1988: Theory, Research and Practice*, pp. 155-170. Center for Social Design, Jacksonville, AL.

Bitgood, S. (2006) An analysis of visitor circulation movement patterns and the general value principle. *Curator: The Museum Journal* 49, 463-475.

Bloomsmith, M.A., Roso, S.R., Bettinger, T., Clay, A.W. and Anderson, U. (2006) Cross-sectional study of the behavioral development of young male chimpanzees in twenty zoos. *American Journal of Primatology* 68, 48-48.

Browning, H. and Maple, T.L. (2019) Developing a metric of usable space for zoo exhibits. *Frontiers in Psychology*. Available at: https://doi.org/10.3389/fpsyg.2019.00791 (accessed 22 June 2021).

Chadwick, C.L., Rees, P.A. and Stevens-Wood, B. (2013) Captive-housed male cheetahs (*Acinonyx jubatus soemmeringeii*) form naturalistic coalitions: measuring associations and calculating chance encounters. *Zoo Biology* 32, 518-527.

Cole, G.C., Naylor, A.D., Hurst, E., Girling, S.J. and Mellanby, R.J. (2020). Hypervitaminosis D in a giant anteater (*Myrmecophaga tridactyla*) and a large hairy armadillo (*Chaetophractus villosus*) receiving a commercial insectivore diet. *Journal of Zoo and Wildlife Medicine* 51, 245-248.

Colman, R.J., Anderson, R.M., Johnson, S.C. *et al.* (2009) Caloric restriction delays disease onset and mortality in rhesus monkeys. *Science* 325, 201-204.

Godinez, A.M. and Fernandez, E.J. (2019) What is the zoo experience? How zoos impact a visitor's behaviours, perceptions, and conservation efforts. *Frontiers in Psychology*. Available at: https://doi.org/10.3389/fpsyg.2019.01746 (accessed 22 June 2021).

Goossens, E., Dorny, P., Boomker, J., Vercammen, F. and Vercruysse, J. (2005) A 12-month survey of the gastrointestinal helminths of antelopes, gazelles and giraffids kept at two zoos in Belgium. *Veterinary Parasitology* 127, 303-312.

Junge, R.E., Gannon, F.H., Porton, I., McAlister, W.H. and Whyte, M.P. (2000) Management and prevention of vitamin D deficiency in captive-born juvenile Chimpanzees (*Pan troglodytes*). *Journal of Zoo and Wildlife Medicine* 32, 361-369.

Moss, A. and Esson, M. (2010) Visitor interest in zoo animals and the implications for collection planning and zoo education programmes. *Zoo Biology* 28, 1-17.

Quirke, T., O'Riordan, R.M. and Zuur, A. (2012) Factors influencing the prevalence of stereotypical behaviour in captive cheetahs (*Acinonyx jubatus*). *Applied Animal Behaviour Science* 142, 189-197.

Rose, P.E., Brereton, J.E., Rowden, L.J., de Figueiredo, R.L. and Riley, L.M. (2019) What's new from the zoo? An analysis of ten years of zoo-themed research output. *Palgrave Communications*, 5, 1-10.

Shepherdson, D.J. (1998) Introduction. Tracing the path of environmental enrichment in zoos. In: Shepherdson, D.J, Mellen, J.D. and Hutchins, M. (eds) *Second Nature: Environmental Enrichment for Captive Animals*, pp. 1-12. Smithsonian Institution Press, Washington, D.C. and London.

Shettel-Neuber, J. and O'Reilly, J. (1981) *Now Where? A Study of Visitor Orientation and Circulation at the Arizona-Senora Desert Museum*. Technical Report No. 87-25. Psychology Institute, Jacksonville State University, Jacksonville, AL.

Weiss, E. and Wilson, S. (2003) The use of classical and operant conditioning in training Aldabra tortoises (*Geochelone gigantea*) for venipuncture and other husbandry issues. *Journal of Applied Animal Welfare Science* 6, 33-38.

Wells, D. (2005) A note on the influence of visitors on the behaviour and welfare of zoo-housed gorillas. *Applied Animal Behaviour Science* 93, 13-17.

Zwinkels, J., Oudegeest, T. and Laterveer, M. (2009) Using visitor observation to evaluate exhibits at the Rotterdam Aquarium. *Visitor Studies* 12, 65-77.